# Computer Science Workbench

Editor: Tosiyasu L. Kunii

W0080376

Brian A. Barsky

# Computer Graphics and Geometric Modeling Using Beta-splines

With 85 Figures, Including 31 in Colour

Springer-Verlag Berlin Heidelberg GmbH

**Prof. Dr. Brian A. Barsky**
Berkeley Computer Graphics Laboratory
Computer Science Division
Department of Electrical Engineering
and Computer Sciences
University of California
Berkeley, CA 94720, USA


*Series editor:*

**Dr. Tosiyasu L. Kunii**
Professor and Director
Kunii Laboratory of Computer Science
Department of Information Science
Faculty of Science, The University of Tokyo
7-3-1 Hongo, Bunkyo-ku
Tokyo 113, Japan

ISBN 978-3-642-72294-3     ISBN 978-3-642-72292-9 (eBook)
DOI 10.1007/978-3-642-72292-9


Library of Congress Cataloging-in-Publication Data
Barsky, Brian A., 1954-
Computer graphics and geometric modeling using Beta-splines.
(Computer science workbench; v. 2)
Bibliography: p.
Includes index.
1. Computer graphics. 2. Spline theory. I. Title. II. Series.
T385.B363  1988  006.6  85-14749
ISBN 978-3-642-72294-3

This work is subject to copyright. All rights are reserved, whether the whole or part of the material is concerned, specifically the rights of translation, reprinting, re-use of illustrations, recitation, broadcasting, reproduction on microfilms or in other ways, and storage in data banks. Duplication of this publication or parts thereof is only permitted under the provisions of the German Copyright Law of September 9, 1965, in its version of June 24, 1985, and a copyright fee must always be paid. Violations fall under the prosecution act of the German Copyright Law.

© Springer-Verlag Berlin Heidelberg 1988
Originally published by Springer-Verlag Berlin Heidelberg New York in 1988
Softcover reprint of the hardcover 1st edition 1988

The use of registered names, trademarks, etc. in this publication does not imply, even in the absence of a specific statement, that such names are exempt from the relevant protective laws and regulations and therefore free for general use.

Typesetting: Asco Trade Typesetting Ltd., Hong Kong; printing: Druckhaus Beltz, Hemsbach/ Bergstr.; binding: J. Schäffer GmbH & Co. KG, Grünstadt
2145/3140-543210

# Preface

In the last decade, it has become apparent that the use of straight line segments and planar polygons to approximate curved lines and surfaces has limited the state-of-the-art in computer graphics. Even with the most sophisticated continuous shading models, polygonal techniques generally result in visually objectionable images. Mach bands are apparent at the borders between adjacent polygons, and there is always a telltale jagged, polygonal silhouette. Also, polygonal methods often require excessive amounts of storage and the "resolution" at which a polygonal database is stored is fixed, independent of the eventual display, as opposed to curved surface techniques in which the resulting image can be computed to whatever level of detail the situation demands.

Early work by Coons introduced the use of nonlinear parametric polynomial representations for the *segments* and *patches* which are stitched together to form *piecewise* curves and surfaces, establishing their viability. More recently, Riesenfeld has advocated the use of B-splines to represent such polynomials on the grounds of greater flexibility and efficiency.

Parametric B-splines have many advantages. Among them is the ability to control the degree of continuity at the joints between adjacent curve segments, and at the borders between surface patches, independent of the order of the segments or the number of control vertices. However, the notion of parametric first or second degree continuity at joints does not always correspond to intuition or to a physically desired effect. For piecewise cubic curves and bicubic surfaces these parametric continuity constraints can be replaced by the more meaningful requirements of continuous unit tangent and curvature vectors. Doing so introduces certain constrained discontinuities in the first and second parametric derivatives. These are expressed in terms of *bias* and *tension* parameters called $\beta 1$ and $\beta 2$, and give rise to *Beta-spline* curves and surfaces.

I am grateful to many people who contributed to this work. Although it is not possible to list them all, there are several people I would like to mention. First, at the University of Utah, I would like express my appreciation to Richard F. Riesenfeld, A. Robin Forrest (School of Computing Studies and Accountancy at the University of East Anglia) and Martin L. Griss for their guidance in this research. Second, at the University of California at Berkeley, I would like to thank Tony D. DeRose, Mark D. Dippé, Kenneth P. Fishkin, and John R. Gross for their interest and assistance in the generation of the shaded images.

Berkeley, California                                                         Brian A. Barsky
October, 1987

In memory of my father
ARTHUR HAROLD BARSKY

# Table of Contents

# 1 Introduction

The underlying concept of this work is the synthesis of two useful concepts: the application of *tension* to a shape; and the study of the *unit tangent vector* and *curvature vector* of a parametrically defined shape as fundamental geometric measures.

Previous techniques of curve and surface representation have been developed from constraints on continuity of parametric derivative vectors. This work proposes replacing these derivatives with the more fundamental geometric measures of the unit tangent vector and curvature vector. This geometric approach has the advantage of adding degrees of freedom which can be captured to provide further control of shape. This control is via *shape parameters*, which generalize previous work on the mathematical modeling of tension.

Tension techniques provide the mathematical methodology to model the application of tension to a shape, a useful tool which provides powerful shape control. The theoretical aspects of two independent methods of applying tension to an *interpolating* spline curve were studied by the author in [1], and both methods were implemented in a pilot curve representation system which was described in [4, 5].

The parametric piecewise representation is explained and various methods of applying tension to a curve are reviewed. Elementary differential geometry concepts are explained and general equations are derived to represent the constraints of continuous unit tangent and curvature vectors. These requirements then replace the conventional conditions of continuous parametric first and second derivative vectors, respectively. These new constraints are expressed in terms of two inherent shape parameters, $\beta 1$ and $\beta 2$, in such a manner that $\beta 1 = 1$ indicates continuity of the parametric first derivative vector and $\beta 1 = 1$ with $\beta 2 = 0$ indicates continuity of the parametric first and second derivative vectors. Since these equations are expressed in terms of the shape parameters, the derivation of the Beta-spline curve representation utilizes symbolic, not numeric, computation. Explicit expressions are provided for the Beta-spline as well as for various derivatives.

Two methods are designed and analyzed for the evaluation and perturbation of a Beta-spline with uniform (assuming a single value over the entire curve) shape parameters. This assumption is then generalized so that the shape parameters are continuous, each varying continuously along the curve, and methods are designed and analyzed for the evaluation and perturbation of the Beta-spline with continuous shape parameters. End conditions for Beta-spline curves are classified and analyzed.

The tensor product Beta-spline surface is then explained. Methods are designed and analyzed for its evaluation and perturbation with uniform shape parameters. The surface representation is generalized for continuous shape parameters, and methods for the evaluation and perturbation in this case are designed and analyzed. Boundary conditions for Beta-spline surfaces are classified and analyzed.

Finally, the geometrical interpretation of these shape parameters and their relation to the mathematical modeling of tension are investigated.

# 2 The Parametric Piecewise Representation

## 2.1 The Parametric Representation

The parametric representation of a curve has each component expressed as a separate univariate (single parameter) function while that of a surface has each component defined by a separate bivariate (two parameter) function. The coordinates of a point can be written as a row vector as follows:

$[X(u) \quad Y(u)]$ for a curve in Euclidean two-space,

$[X(u) \quad Y(u) \quad Z(u)]$ for a curve in Euclidean three-space, and

$[X(u, v) \quad Y(u, v) \quad Z(u, v)]$ for a surface in Euclidean three-space.

For notational convenience, the row vectors for a curve and surface will be denoted as $\mathbf{Q}(u)$ and $\mathbf{Q}(u, v)$, respectively.

A parametric derivative, with respect to some parameter or parameters, can also be represented as a row vector. Each component is the derivative, with respect to that parameter or parameters, of the function corresponding to its coordinate. These parametric derivative vectors are then:

$$\frac{d^a}{du^a} \mathbf{Q}(u) = \left[ \frac{d^a}{du^a} X(u) \quad \frac{d^a}{du^a} Y(u) \right]$$

for a curve in Euclidean two-space,

$$\frac{d^a}{du^a} \mathbf{Q}(u) = \left[ \frac{d^a}{du^a} X(u) \quad \frac{d^a}{du^a} Y(u) \quad \frac{d^a}{du^a} Z(u) \right]$$

for a curve in Euclidean three-space, and

$$\frac{\partial^{a+b}}{\partial u^a \partial v^b} \mathbf{Q}(u, v) = \left[ \frac{\partial^{a+b}}{\partial u^a \partial v^b} X(u, v) \quad \frac{\partial^{a+b}}{\partial u^a \partial v^b} Y(u, v) \quad \frac{\partial^{a+b}}{\partial u^a \partial v^b} Z(u, v) \right]$$

for a surface in Euclidean three-space.

For notational convenience define

$$Q^{(a)}(c) = \frac{d^a}{du^a} Q(u)\Big|_{u=c}$$

$$\qquad\qquad\qquad\qquad\qquad\qquad\qquad\qquad (2.1)$$

$$Q^{(a,b)}(c, d) = \frac{\partial^{a+b}}{\partial u^a \partial v^b} Q(u, v)\Big|_{u=c, v=d} \quad.$$

## 2.2 The Piecewise Representation

Although an entire curve or surface is not easily defined by a single analytic function, it can be apportioned into a set of smaller pieces, each described by a separate analytic function, to form a *piecewise representation*. A spline curve is composed of a sequence of polynomials called spline *curve segments* (Figure 2.1) while a spline surface is a mosaic of *surface patches* (Figure 2.2).

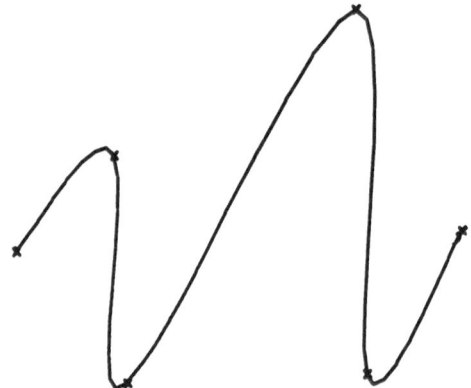

**Fig. 2.1.**   A spline curve is a sequence of curve segments.

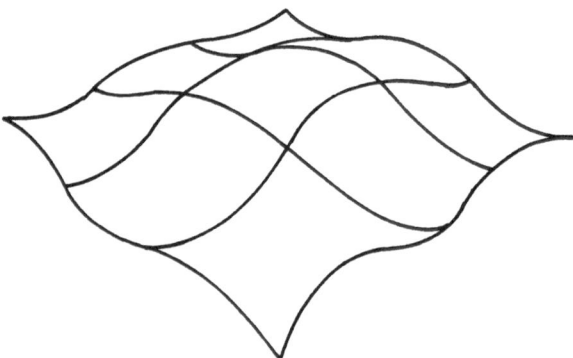

**Fig. 2.2.**   A spline surface is a mosaic of surface patches.

# 3 The Application of Tension to a Curve

## 3.1 Spline Under Tension

### 3.1.1 Explanation

The cubic spline sometimes exhibits unnecessary oscillations due to "extraneous" inflection points. In order to eliminate them, it is desirable to intuitively "pull out" these points by increasing tension. This concept was first analytically modeled by Schweikert in [23] and an alternative development was given in [6] and generalized in [19]. A detailed derivation of the generalized form based on a variational principle is given in [1].

Continuity of position and the second derivative vector are trivially achieved by their specification at each interior point and continuity of the first derivative vector is achieved by solving a system of equations for the set of values of the second derivative vector. In particular, let $\{P_0, P_1, \ldots, P_m\}$ be a set of $m + 1$ points to be interpolated and $\{P_0^2, P_1^2, \ldots, P_m^2\}$ be the set of corresponding values of the second derivative vector. Denoting the tension over the $i^{\text{th}}$ curve segment as $s_i$, the $i^{\text{th}}$ curve segment can be written as

$$Q_i(u) = \frac{P_{i-1}^2 \sinh(s_i(u_i - u)) + P_i^2 \sinh(s_i(u - u_{i-1}))}{s_i^2 \sinh(s_i \Delta u_i)}$$

$$+ \left(P_{i-1} - \frac{P_{i-1}^2}{s_i^2}\right)\frac{u_i - u}{\Delta u_i} + \left(P_i - \frac{P_i^2}{s_i^2}\right)\frac{u - u_{i-1}}{\Delta u_i} \qquad (3.1)$$

where

$$\Delta u_i = u_i - u_{i-1} \qquad \qquad \text{for } u_{i-1} \leqslant u < u_i \quad \text{for } i = 1, 2, \ldots, m .$$

### 3.1.2 Determining the Set of Values of the Second Derivative Vector

The continuity requirement of the first derivative vector at the interior point $P_i$ is represented by the equation

$$l_i P_{i-1}^2 + (n_i + n_{i+1})P_i^2 + l_{i+1} P_{i+1}^2 = R_{i+1} - R_i \qquad (3.2)$$

where

$$l_i = \frac{\dfrac{1}{v_i} - \operatorname{csch}(v_i)}{s_i}$$

$$n_i = \frac{\operatorname{coth}(v_i) - \dfrac{1}{v_i}}{s_i}$$

$$\mathbf{R}_i = \frac{\mathbf{P}_i - \mathbf{P}_{i-1}}{\Delta u_i}$$

and where

$$v_i = s_i \Delta u_i \qquad\qquad\qquad \text{for } i = 1, 2, \ldots, m - 1 \ .$$

This is a system of $m - 1$ linear equations for the $m + 1$ unknown values of the second derivative vector. To find a unique solution, two more equations are needed, and this can be fulfilled by specifying an end condition at both $\mathbf{P}_0$ and $\mathbf{P}_m$. The complete system of equations for both the first derivative vector specification end condition and the natural spline end condition are given in [1].

## 3.2 v-Spline

### 3.2.1 Explanation

An objection to the spline under tension is that it is expressed in terms of exponential functions rather than polynomials, which is a major impediment to efficient evaluation. To circumvent this problem, a polynomial alternative to the spline under tension was developed by Nielson in [17, 18] which he called the v-spline. It is derived in detail in [1] using the cubic Hermite basis functions approach, thereby emphasizing its relation to the conventional cubic interpolatory spline.

It is important to note that each tension value for a v-spline is associated with a *point* to be interpolated, not a spline curve *segment* as is the case with a spline under tension. A curve segment does, however, converge to a straight line segment as the tension values at both endpoints are increased. In addition, the number of tension values for a v-spline is therefore one more than that for a spline under tension.

Unlike the spline under tension, the v-spline does not have continuity of the second parametric derivative vector when the tension values are nonzero. The discontinuity in the second parametric derivative vector at an interior point to be interpolated is parallel to the corresponding first derivative vector, and the ratio of the respective magnitudes is the absolute value of the tension value there. Specifically,

$$\mathbf{Q}_{i+1}^{(2)}(u_i) - \mathbf{Q}_i^{(2)}(u_i) = t_i \mathbf{Q}_{i+1}^{(1)}(u_i) = t_i \mathbf{Q}_i^{(1)}(u_i) \qquad\qquad (3.3)$$

where

$$t_i \text{ is the tension value on the point } \mathbf{P}_i \qquad\qquad \text{for } i = 1, 2, \ldots, m - 1 \ .$$

Note that the magnitude of this discontinuity reduces to zero for a zero tension value, corresponding to the continuous second parametric derivative vector of the cubic interpolatory spline. However, the $v$-spline does have geometric second derivative continuity; that is, $\dfrac{d^2 y_Q}{dx_Q^2}$ is continuous where $x_Q^{(1)}(u)$ is nonzero, and similarly $\dfrac{d^2 x_Q}{dy_Q^2}$ is continuous where $y_Q^{(1)}(u)$ is nonzero. This implies that the *curvature* of a $v$-spline is also continuous [17] (see Section 5.2).

### 3.2.2 Determining the Set of Values of the First Derivative Vector

Since each $v$-spline curve segment is a cubic polynomial, it can be represented by specifying the position of each endpoint along with the corresponding value of the first derivative vector. This specifies *cubic Hermite interpolation* [22] which has continuity of position and the first derivative vector trivially guaranteed by their specification at each interior point.

Specifically, let $\{P_0, P_1, \ldots, P_m\}$ be a set of $m + 1$ points to be interpolated and $\{P_0^1, P_1^1, \ldots, P_m^1\}$ be the set of corresponding values of the first derivative vector. Then the $i^{th}$ curve segment can be written as

$$Q_i(u) = \sum_{j=0}^{1} \sum_{k=0}^{1} g_{j,k}(u_{i-1}, u_i; u) * P_{i-1+k}^j \qquad (3.4)$$

$$\text{for } u_{i-1} \leqslant u < u_i \quad \text{for } i = 1, 2, \ldots, m \ .$$

The functions $g_{j,k}(u_{i-1}, u_i; u)$ are the *generalized cubic Hermite basis functions* and can be written in matrix form as

$$[g_{0,0}(u_{i-1}, u_i; u) \quad g_{0,1}(u_{i-1}, u_i; u) \quad g_{1,0}(u_{i-1}, u_i; u) \quad g_{1,1}(u_{i-1}, u_i; u)]$$

$$= [w^3 \quad w^2 \quad w \quad 1] \begin{bmatrix} 2 & -2 & 1 & 1 \\ -3 & 3 & -2 & -1 \\ 0 & 0 & 1 & 0 \\ 1 & 0 & 0 & 0 \end{bmatrix} \begin{bmatrix} 1 & 0 & 0 & 0 \\ 0 & 1 & 0 & 0 \\ 0 & 0 & \Delta u_i & 0 \\ 0 & 0 & 0 & \Delta u_i \end{bmatrix} \qquad (3.5)$$

where

$$w = \frac{u - u_{i-1}}{\Delta u_i} \quad \text{and} \quad \Delta u_i = u_i - u_{i-1} \ .$$

There exists a special set of values of the first derivative vector with the property that the resulting piecewise cubic Hermite curve is a $v$-spline. Thus, a $v$-spline can be constructed by first determining this set of values and then invoking cubic Hermite interpolation using these values of the first derivative vector at the interior points. The process of determining this set of values requires the formulation and subsequent solution of a system of simultaneous linear equations.

Although the $v$-spline does not generally have continuity of the second parametric derivative vector, the amount of discontinuity at an interior point to be interpolated is given by equation (3.3); therefore, this equation can be used to determine the values of the first derivative at the interior points. Differentiating the expression for the $i^{th}$ cubic Hermite curve segment given in equations (3.4) and (3.5) twice with respect to $u$ results in

$$Q_i^{(2)}(u) = \sum_{j=0}^{1} \sum_{k=0}^{1} g_{j,k}^{(2)}(u_{i-1}, u_i; u) * P_{i-1+k}^j \tag{3.6}$$

$$\text{for } u_{i-1} \leqslant u < u_i \quad \text{for } i = 1, 2, \ldots, m ,$$

where the functions $g_{j,k}^{(2)}(u_{i-1}, u_i; u)$ can be written in matrix form as

$$[g_{0,0}^{(2)}(u_{i-1}, u_i; u) \quad g_{0,1}^{(2)}(u_{i-1}, u_i; u) \quad g_{1,0}^{(2)}(u_{i-1}, u_i; u) \quad g_{1,1}^{(2)}(u_{i-1}, u_i; u)]$$

$$= \left[ \frac{u - u_{i-1}}{\Delta u_i} \quad 1 \right] \begin{bmatrix} 2 & -2 & 3 & 3 \\ -1 & 1 & -2 & -1 \end{bmatrix} \begin{bmatrix} \dfrac{6}{\Delta^2 u_i} & 0 & 0 & 0 \\ 0 & \dfrac{6}{\Delta^2 u_i} & 0 & 0 \\ 0 & 0 & \dfrac{2}{\Delta u_i} & 0 \\ 0 & 0 & 0 & \dfrac{2}{\Delta u_i} \end{bmatrix} \tag{3.7}$$

and

$$\Delta u_i = u_i - u_{i-1} .$$

Using equations (3.6) and (3.7) to evaluate both terms on the left-hand side of equation (3.3), and performing some algebraic simplification results in the following equation:

$$\Delta u_{i+1} P_{i-1}^1 + \left[ 2(\Delta u_i + \Delta u_{i+1}) + \frac{t_i}{2} \right] P_i^1 + \Delta u_i P_{i+1}^1$$

$$= 3 \left[ \frac{\Delta u_{i+1}}{\Delta u_i} (P_i - P_{i-1}) + \frac{\Delta u_i}{\Delta u_{i+1}} (P_{i+1} - P_i) \right] \quad \text{for } i = 1, 2, \ldots, m - 1 . \tag{3.8}$$

This is a system of $m - 1$ equations for the $m + 1$ unknown values of the first derivative vector. To find a unique solution, two more equations are needed, and this can be fulfilled by specifying an end condition at both $P_0$ and $P_m$. The complete system of equations for two different end conditions is provided in [1].

## 3.3 Tension Methods with Local Control

Unfortunately, none of the above-mentioned interpolating spline curve representations provides *local control*; that is, the capability of modifying one portion of the curve without altering the remainder of the curve. Local control is inherent in the B-spline formulation [2, 20], and thus this would be a good starting representation upon which to apply tension. In an approach that preserved the variation diminishing property of the B-spline scheme [2], Lane experimented with this idea by adding knots to a nonuniform B-spline curve in the region of desired tension [16].

The concept which is to be developed herein is the derivation of a new curve/ surface representation having an inherent capability to model tension. This representation will be called the *Beta-spline*, and will contain the uniform cubic B-spline as a special case. The additional flexibility of this tension capability is accomplished through careful examination of the unit tangent vector and curvature vector of a parametrically defined shape.

# 4 Elementary Differential Geometry Concepts

## 4.1 Unit Tangent Vector

Consider a space curve (in three dimensions) parametrized with respect to an arbitrary parameter $u$ [8, 9, 10, 15, 24]. The unit tangent vector has the same direction and sense as the parametric first derivative vector, but it is normalized. Denoting it by $\mathbf{T}(u)$, and assuming a nonzero first derivative,

$$\mathbf{T}(u) = \frac{\mathbf{Q}^{(1)}(u)}{|\mathbf{Q}^{(1)}(u)|} \ . \tag{4.1}$$

In the special case where the curve is parametrized according to arc length $s$,

$$\frac{d}{ds}\mathbf{Q} = \mathbf{T}(s) \ . \tag{4.2}$$

To prove this, form

$$\frac{d}{ds}\mathbf{Q} = \frac{\mathbf{Q}^{(1)}(u)}{s^{(1)}(u)} \ . \tag{4.3}$$

Since $s(u)$ is arc length,

$$s(u) = \int_a^u |\mathbf{Q}^{(1)}(t)| \, dt \ . \tag{4.4}$$

Therefore,

$$s^{(1)}(u) = |\mathbf{Q}^{(1)}(u)| \ . \tag{4.5}$$

Thus the first derivative of the arc length with respect to the parameter $u$ is the length of the first derivative vector of the curve with respect to the same parameter. Substituting equation (4.5) into equation (4.3) yields the same right-hand side as equation (4.1); thus,

$$\frac{d}{ds}\mathbf{Q} = \mathbf{T}(s) \ . \tag{4.6}$$

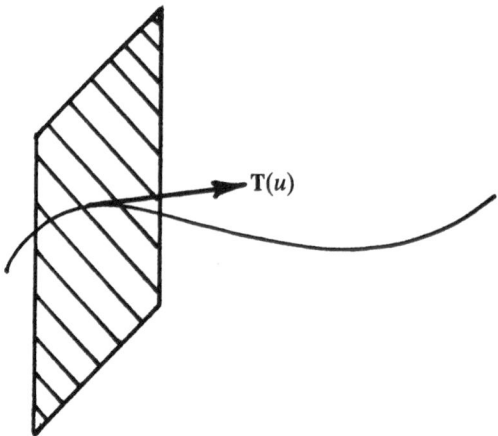

**Fig. 4.1.** The normal plane.

## 4.2 The Principal Normal

At a given point on a space curve there are an infinite number of normal vectors which together sweep out a plane known as the *normal plane* (Figure 4.1).

Consider now the following lemma which shows that at a given parametric value, a vector-valued function having constant magnitude is orthogonal to its parametric derivative vector.

**Lemma.** Let $\mathbf{F}: \mathbb{R} \to \mathbb{R}^n$ be a differentiable function having constant magnitude $m$. Then $\mathbf{F}(u)$ is orthogonal to $\mathbf{F}^{(1)}(u)$.

**Proof.** Form $|\mathbf{F}(u)| = m$. Thus

$$\mathbf{F}(u) \cdot \mathbf{F}(u) = m^2 \ .$$

Differentiating with respect to $u$,

$$\mathbf{F}(u) \cdot \mathbf{F}^{(1)}(u) + \mathbf{F}^{(1)}(u) \cdot \mathbf{F}(u) = 0 \ .$$

By the commutativity of the dot product,

$$\mathbf{F}(u) \cdot \mathbf{F}^{(1)}(u) = 0 \ .$$

Hence $\mathbf{F}(u)$ is orthogonal to $\mathbf{F}^{(1)}(u)$.

Since $\mathbf{T}(u)$ is a unit vector, the above lemma can be invoked to show that $\mathbf{T}^{(1)}(u)$ is orthogonal to $\mathbf{T}(u)$. Normalizing $\mathbf{T}^{(1)}(u)$ yields the *principal normal vector* denoted by $\mathbf{N}(u)$:

$$\mathbf{N}(u) = \frac{\mathbf{T}^{(1)}(u)}{|\mathbf{T}^{(1)}(u)|} \ . \tag{4.7}$$

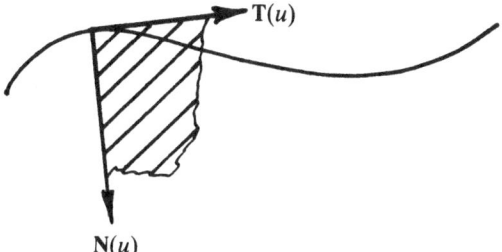

**Fig. 4.2.** The osculating plane.

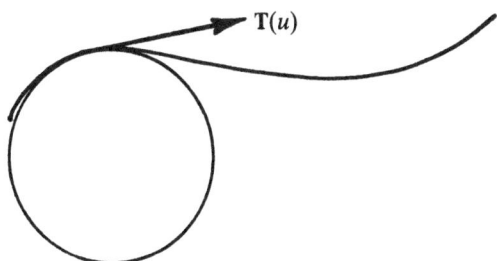

**Fig. 4.3.** The osculating circle.

## 4.3 The Osculating Plane and Osculating Circle

The plane containing the vectors $T(u)$ and $N(u)$ at a given point on the curve is called the *osculating plane* (Figure 4.2) because it is the plane which has the *greatest contact* with the curve.

The *osculating circle* at a given point on the curve is the circle which most closely approximates the curve at that point (Figure 4.3). More formally, it is the circle whose first and second derivative vectors agree with those of the curve at the given point, and which lies on the concave side of the curve. Now, the osculating circle lies in the osculating plane. To show this, the normal to the plane of the osculating circle is compared to that of the osculating plane. Applying the chain rule to equation (4.6) yields

$$\frac{Q^{(1)}(u)}{s^{(1)}(u)} = T(u) \ . \tag{4.8}$$

Then

$$Q^{(1)}(u) = T(u)s^{(1)}(u) \ . \tag{4.9}$$

Differentiating with respect to $u$,

$$Q^{(2)}(u) = T(u)s^{(2)}(u) + T^{(1)}(u)s^{(1)}(u) \ . \tag{4.10}$$

Using equations (4.9) and (4.10), and noting that $T(u) \times T(u) = 0$, the (scaled) normal to the plane of the osculating circle is then given by

$$Q^{(1)}(u) \times Q^{(2)}(u) = [s^{(1)}(u)]^2 T(u) \times T^{(1)}(u) \ . \tag{4.11}$$

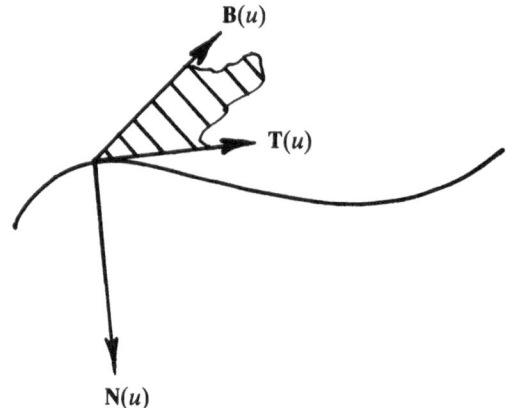

**Fig. 4.4.** The rectifying plane.

From equation (4.7)

$$\mathbf{T}^{(1)}(u) = \mathbf{N}(u)|\mathbf{T}^{(1)}(u)| \ .$$
(4.12)

Thus,

$$\mathbf{Q}^{(1)}(u) \times \mathbf{Q}^{(2)}(u) = [s^{(1)}(u)]^2 |\mathbf{T}^{(1)}(u)| \mathbf{T}(u) \times \mathbf{N}(u) \ .$$
(4.13)

Since the unit normal to the osculating plane is $\mathbf{T}(u) \times \mathbf{N}(u)$, the (scaled) normal to the plane of the osculating circle is in the same direction as that of the osculating plane.

## 4.4  The Binormal

For notational convenience, the unit normal to the osculating plane is termed the *binormal vector* $\mathbf{B}(u)$; that is,

$$\mathbf{B}(u) = \mathbf{T}(u) \times \mathbf{N}(u) \ .$$
(4.14)

At a given point on the curve the *rectifying plane* contains the vectors $\mathbf{B}(u)$ and $\mathbf{T}(u)$ and has the normal $\mathbf{N}(u)$ (Figure 4.4).

The three unit vectors $\langle \mathbf{T}(u), \mathbf{N}(u), \mathbf{B}(u) \rangle$ form a right-handed set of mutually orthogonal vectors called the *moving trihedron* of the curve which is an intrinsic coordinate system.

## 4.5  Curvature and Curvature Vector

The center and radius of the osculating circle are called the *center of curvature* $\mathbf{c}(u)$ and *radius of curvature* $\rho(u)$, respectively, at this point.

The *curvature* $\kappa(u)$, at a point, is defined to be the reciprocal of the radius of curvature. The *curvature vector* $\mathbf{K}(u)$ has a magnitude equal to the curvature and it points from the given point to the center of curvature; that is, in the direction of

$\mathbf{T}^{(1)}(u)$. Therefore, it is given by

$$\mathbf{K}(u) = \kappa(u)\mathbf{N}(u) .$$ (4.15)

### 4.5.1 Arc Length Parametrization

The curvature vector is equivalent to the second derivative in the case where this derivative is with respect to arc length; that is,

$$\mathbf{K}(s) = \frac{d^2}{ds^2}\mathbf{Q}$$ (4.16)

where

$\mathbf{K}(s)$ is the curvature vector and $s$ is arc length.

This can be proven by showing that both sides of equation (4.16) are identical in magnitude, direction, and sense.

From Figure 4.5,

$$ds = \rho(s)\,d\theta .$$ (4.17)

Therefore

$$\kappa(s) = \frac{1}{\rho(s)} = \frac{d\theta}{ds} .$$ (4.18)

Considering Figure 4.6, since the unit tangent vectors are of unit length,

$$d\theta = \left| \frac{d}{ds}\mathbf{Q}(s+ds) - \frac{d}{ds}\mathbf{Q}(s) \right|$$

$$d\theta = \left| ds\, \frac{d^2}{ds^2}\mathbf{Q} \right|$$

$$\frac{d\theta}{ds} = \left| \frac{d^2}{ds^2}\mathbf{Q} \right| .$$ (4.19)

From equations (4.15), (4.18), and (4.19),

$$\kappa(s) = \left| \frac{d^2}{ds^2}\mathbf{Q} \right|$$ (4.20)

and thus the magnitudes are the same.

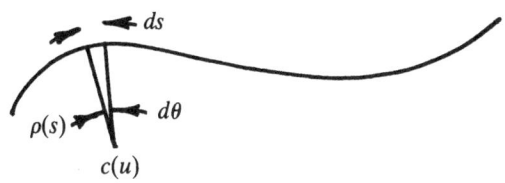

**Fig. 4.5.** The center of curvature and radius of curvature.

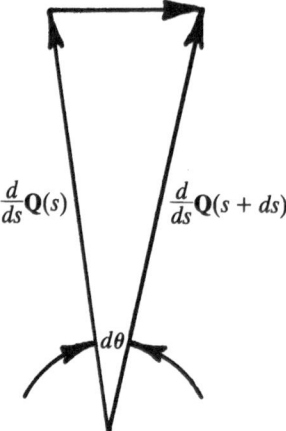

**Fig. 4.6.** Difference of unit tangent vectors.

To show the equivalence of the directions, recall from Section 4.1 that $\mathbf{T}(s)$ is a unit vector. Now the lemma in Section 4.2 can be invoked to show that

$$\mathbf{T}(s) \cdot \frac{d}{ds} \mathbf{T} = 0 \ . \tag{4.21}$$

Recalling equation (4.6)

$$\mathbf{T}(s) \cdot \frac{d^2}{ds^2} \mathbf{Q} = 0 \tag{4.22}$$

and thus $\dfrac{d^2}{ds^2} \mathbf{Q}$ is perpendicular to $\mathbf{T}(s)$. Now, the curvature vector has the same direction as the radius of curvature which lies in the osculating plane and is perpendicular to the unit tangent vector $\mathbf{T}(s)$. Thus $\mathbf{K}(s)$ and $\dfrac{d^2}{ds^2} \mathbf{Q}$ are both perpendicular to $\mathbf{T}(s)$ and all three vectors lie in the osculating plane; hence, $\mathbf{K}(s)$ and $\dfrac{d^2}{ds^2} \mathbf{Q}$ must have the same direction.

Finally, it can be seen that the change in the unit tangent vector is towards the center of curvature, and hence $\dfrac{d^2}{ds^2} \mathbf{Q}$ and $\mathbf{K}(s)$ have the same sense.

For a planar curve, it is the usual convention to assign a positive curvature to a curve curving "to the left" while a negative curvature denotes a curve curving "to the right" (Figure 4.7).

### 4.5.2  Arbitrary Parametrization – Geometric Form

In the computer aided geometric design field, arc length parametrization of a curve cannot be assumed because the curve itself does not yet exist. Since the curve is not parametrized with respect to arc length, the formulae of the preceding section are

$$(a)$$
$$+$$

$$(b)$$

**Fig. 4.7.** Sign of curvature. (a) A planar curve curving to the left has positive curvature. (b) A planar curve curving to the right has negative curvature.

difficult to apply. The curvature vector for a curve parametrized with respect to an arbitrary parameter $u$ will now be derived.

Recalling equation (4.16),

$$
\begin{aligned}
\mathbf{K}(s) &= \frac{d}{ds}\frac{d}{ds}\mathbf{Q} \\[2mm]
&= \frac{d}{ds}\frac{\mathbf{Q}^{(1)}(u)}{s^{(1)}(u)} \\[2mm]
&= \frac{d}{du}\frac{\dfrac{\mathbf{Q}^{(1)}(u)}{s^{(1)}(u)}}{s^{(1)}(u)} \\[2mm]
&= \frac{s^{(1)}(u)\mathbf{Q}^{(2)}(u) - \mathbf{Q}^{(1)}(u)s^{(2)}(u)}{[s^{(1)}(u)]^3} \; .
\end{aligned}
\tag{4.23}
$$

Differentiating equation (4.5),

$$
\begin{aligned}
s^{(2)}(u) &= \frac{d}{du}|\mathbf{Q}^{(1)}(u)| \\[2mm]
&= \frac{d}{du}[\mathbf{Q}^{(1)}(u)\cdot\mathbf{Q}^{(1)}(u)]^{1/2} \\[2mm]
&= \frac{\mathbf{Q}^{(1)}(u)\cdot\mathbf{Q}^{(2)}(u) + \mathbf{Q}^{(2)}(u)\cdot\mathbf{Q}^{(1)}(u)}{2[\mathbf{Q}^{(1)}(u)\cdot\mathbf{Q}^{(1)}(u)]^{1/2}} \\[2mm]
&= \frac{\mathbf{Q}^{(1)}(u)\cdot\mathbf{Q}^{(2)}(u)}{[\mathbf{Q}^{(1)}(u)\cdot\mathbf{Q}^{(1)}(u)]^{1/2}} \\[2mm]
&= \frac{\mathbf{Q}^{(2)}(u)\cdot\mathbf{Q}^{(1)}(u)}{|\mathbf{Q}^{(1)}(u)|} \; .
\end{aligned}
\tag{4.24}
$$

Recalling equations (4.5) and (4.1), it can be seen that the rate of change of the length of the parametric first derivative vector is equal to the component of the parametric second derivative vector in the direction of the unit tangent vector. Again using equation (4.5),

$$
s^{(2)}(u) = \mathbf{Q}^{(2)}(u)\cdot\frac{\mathbf{Q}^{(1)}(u)}{s^{(1)}(u)} \; .
\tag{4.25}
$$

Substituting equation (4.25) into equation (4.23) yields

$$K(s) = \frac{\mathbf{Q}^{(2)}(u) - \dfrac{\mathbf{Q}^{(1)}(u)}{s^{(1)}(u)}\left[\mathbf{Q}^{(2)}(u) \cdot \dfrac{\mathbf{Q}^{(1)}(u)}{s^{(1)}(u)}\right]}{[s^{(1)}(u)]^2}. \tag{4.26}$$

Therefore, from equations (4.5) and (4.8),

$$K(u) = \frac{\mathbf{Q}^{(2)}(u) - \mathbf{T}(u)[\mathbf{Q}^{(2)}(u) \cdot \mathbf{T}(u)]}{|\mathbf{Q}^{(1)}(u)|^2}. \tag{4.27}$$

Since the curvature vector $\mathbf{K}(u)$, the unit tangent vector $\mathbf{T}(u)$, and the parametric second derivative vector $\mathbf{Q}^{(2)}(u)$ all lie in the osculating plane, the graphical interpretation of the curvature vector is the scaled component of the second derivative vector in the direction perpendicular to the unit tangent vector (Figure 4.8).

Since the curvature is the magnitude of the curvature vector,

$$\kappa(u) = |\mathbf{K}(u)| = \frac{|\mathbf{Q}^{(2)}(u) - \mathbf{T}(u)[\mathbf{Q}^{(2)}(u) \cdot \mathbf{T}(u)]|}{|\mathbf{Q}^{(1)}(u)|^2}. \tag{4.28}$$

### 4.5.3 Arbitrary Parametrization – Cross-product Form

Considering Figure 4.8, standard trigonometric identities show that the numerator of equation (4.28) is

$$|\mathbf{Q}^{(2)}(u)| \sin \theta \tag{4.29}$$

where $\theta$ is the angle between the unit tangent vector and the second derivative vector.

From this, an alternative *cross-product* formulation can be derived for curvature. Since $\mathbf{T}(u)$ is of unit length, expression (4.29) can be written as

$$|\mathbf{T}(u)||\mathbf{Q}^{(2)}(u)| \sin \theta \tag{4.30}$$

which, by definition of the cross product, is

$$|\mathbf{T}(u) \times \mathbf{Q}^{(2)}(u)|. \tag{4.31}$$

Substituting this expression for the numerator of equation (4.28) yields

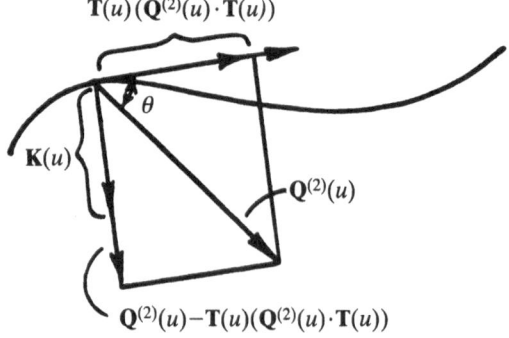

**Fig. 4.8.** Graphical interpretation of the curvature vector formula.

$$\kappa(u) = \frac{|\mathbf{T}(u) \times \mathbf{Q}^{(2)}(u)|}{|\mathbf{Q}^{(1)}(u)|^2} \tag{4.32}$$

and from equation (4.1),

$$\kappa(u) = \frac{|\mathbf{Q}^{(1)}(u) \times \mathbf{Q}^{(2)}(u)|}{|\mathbf{Q}^{(1)}(u)|^3} . \tag{4.33}$$

A cross-product form can also be derived for the curvature vector. From equation (4.15),

$$\mathbf{K}(u) = \kappa(u)\mathbf{B}(u) \times \mathbf{T}(u) . \tag{4.34}$$

Since $\mathbf{B}(u)$ is normal to the osculating plane which contains $\mathbf{Q}^{(1)}(u)$ and $\mathbf{Q}^{(2)}(u)$, and is of unit length,

$$\mathbf{B}(u) = \frac{\mathbf{Q}^{(1)}(u) \times \mathbf{Q}^{(2)}(u)}{|\mathbf{Q}^{(1)}(u) \times \mathbf{Q}^{(2)}(u)|} . \tag{4.35}$$

Combining equations (4.1), (4.34), and (4.35),

$$\mathbf{K}(u) = \frac{(\mathbf{Q}^{(1)}(u) \times \mathbf{Q}^{(2)}(u)) \times \mathbf{Q}^{(1)}(u)}{|\mathbf{Q}^{(1)}(u)|^4} . \tag{4.36}$$

### 4.5.4 Conditions for Zero Curvature Vector

Note that one cannot conclude that the curvature vector has vanished solely on the basis of the value of the second derivative vector. Two sufficient conditions along with their corresponding conclusions are:

1. If the first derivative vector is nonzero, and the first and second derivative vectors are linearly dependent, then the curvature vector is zero.
2. If the first and second derivative vectors are linearly independent, then the curvature vector is nonzero.

Since linearly independent vectors must both be nonzero, a zero first derivative vector does not satisfy either of these conditions; thus, further analysis is required in order to reach a conclusion about the curvature vector in this case.

# 5 Fundamental Geometric Measures

## 5.1 Continuity

In the field of computer aided geometric design, an important property of each of the various mathematical techniques of shape representation is its degree of *continuity*; that is, the highest level of differentiation which is continuous. More formally:

If a function $f \in C^n[a, b]$

then $f^{(i)}$, $i = 0, 1, \ldots, n$, exist and are continuous over $[a, b]$ .

$$(5.1)$$

Since these techniques utilize a *parametric representation* [2], continuity is usually expressed in terms of parametric derivatives, which are represented as row vectors as was explained in Section 2.1.

## 5.2 The Unit Tangent Vector and Curvature Vector

However, it is this author's contention that parametric derivative vectors do not provide an appropriate measure of continuity in the context of geometric design and modeling. More appropriate geometric measures would be the unit tangent vector and curvature vector, instead of parametric first and second derivative vectors, respectively. Although these terms are frequently used interchangeably in the literature, this is due to carelessness, not equivalence. Some examples illustrating the differences follow.

Figure 5.1 shows the two straight line segments given by

$$Q_1(u) = [12u, \ 9u] \qquad\qquad 0 \leqslant u \leqslant 1$$
$$Q_2(u) = [4(u + 3), \ 3(u + 3)] \qquad\qquad 0 \leqslant u \leqslant 1 \ .$$

$$(5.2)$$

The parametric first derivative vectors are

$$Q_1^{(1)}(u) = [12, \ 9]$$
$$Q_2^{(1)}(u) = [4, \ 3] \ .$$

$$(5.3)$$

Even though the segments join with a discontinuous parametric first derivative

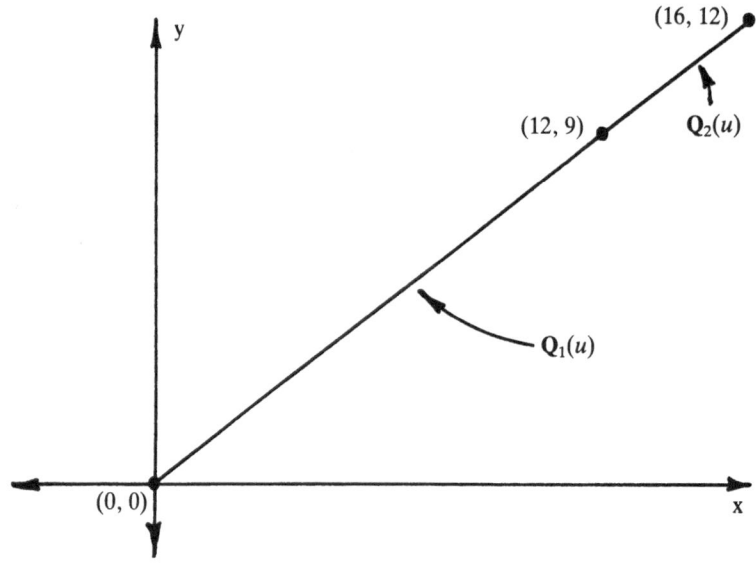

**Fig. 5.1.** Discontinuous first derivative yet continuous unit tangent.

vector, the unit tangent vector is continuous at the joint; specifically, $\mathbf{T} = \left[\dfrac{4}{5}, \dfrac{3}{5}\right]$.

Now consider the two straight line segments shown in Figure 5.2:

$$\mathbf{Q}_3(u) = [a(2u - u^2),\ b(2u - u^2)] \qquad\qquad 0 \leqslant u \leqslant 1$$
$$\mathbf{Q}_4(u) = [a + (c - a)u^2,\ b(1 - u^2)] \qquad\qquad 0 \leqslant u \leqslant 1 .\qquad (5.4)$$

The parametric first derivative vectors are

$$\mathbf{Q}_3^{(1)}(u) = [a(2 - 2u),\ b(2 - 2u)] \qquad\qquad 0 \leqslant u \leqslant 1$$
$$\mathbf{Q}_4^{(1)}(u) = [2(c - a)u,\ -2bu] \qquad\qquad 0 \leqslant u \leqslant 1 .\qquad (5.5)$$

Evaluating equation (5.5) at the joint yields

$$\mathbf{Q}_3^{(1)}(1) = [0,\ 0]$$
$$\mathbf{Q}_4^{(1)}(0) = [0,\ 0] .\qquad (5.6)$$

Thus the segments join with a continuous parametric first derivative vector even though the unit tangent vector is clearly discontinuous at the joint.

Figure 5.3 shows a semi-circle of radius $a$ formed by two quarter-circle segments, parametrized as follows:

$$\mathbf{Q}_5(u) = \left[ a \sin\left(\frac{\pi}{2}u^2\right),\ a \cos\left(\frac{\pi}{2}u^2\right) \right] \qquad\qquad 0 \leqslant u \leqslant 1$$
$$\mathbf{Q}_6(u) = \left[ a \cos\left(\frac{\pi}{2}u^2\right),\ -a \sin\left(\frac{\pi}{2}u^2\right) \right] \qquad\qquad 0 \leqslant u \leqslant 1 .\qquad (5.7)$$

The parametric second derivative vectors are

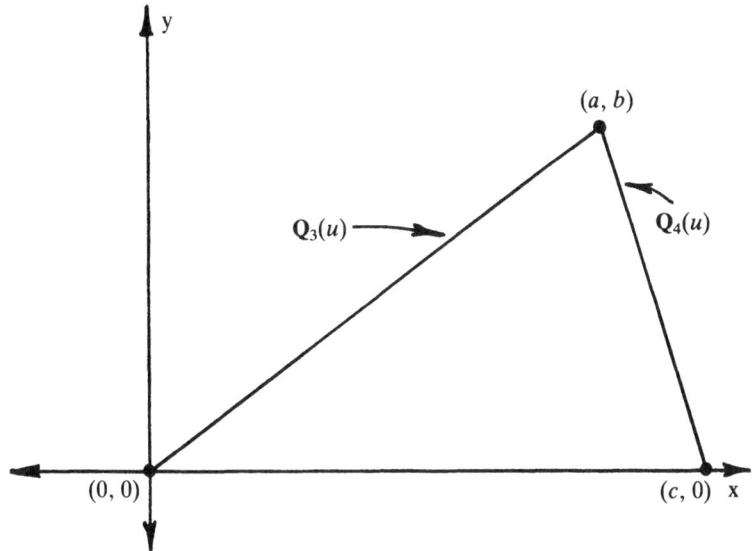

**Fig. 5.2.** Continuous first derivative yet discontinuous unit tangent.

$$Q_5^{(2)}(u) = \left[ a\pi \cos\left(\frac{\pi}{2}u^2\right) - a(\pi u)^2 \sin\left(\frac{\pi}{2}u^2\right), \right.$$

$$\left. - a\pi \sin\left(\frac{\pi}{2}u^2\right) - a(\pi u)^2 \cos\left(\frac{\pi}{2}u^2\right) \right] \qquad 0 \leqslant u \leqslant 1$$

$$(5.8)$$

$$Q_6^{(2)}(u) = \left[ -a\pi \sin\left(\frac{\pi}{2}u^2\right) - a(\pi u)^2 \cos\left(\frac{\pi}{2}u^2\right), \right.$$

$$\left. - a\pi \cos\left(\frac{\pi}{2}u^2\right) + a(\pi u)^2 \sin\left(\frac{\pi}{2}u^2\right) \right] \qquad 0 \leqslant u \leqslant 1 .$$

The values of these derivatives (equation (5.8)) at the joint are

$$Q_5^{(2)}(1) = [-a\pi^2, \ -a\pi]$$

$$Q_6^{(2)}(0) = [0, \ -a\pi] .$$

$$(5.9)$$

Although the segments join with a discontinuous parametric second derivative vector, the curvature vector is clearly continuous, pointing from $(a, 0)$ towards the origin with a magnitude of $1/a$, since this is a circular arc of radius $a$.

In Figure 5.4, two quarter-circle segments of radii $a$ and $b$, respectively, are joined:

$$Q_7(u) = \left[ a \cos\left(\frac{\pi}{2}(1-u)^3\right), \ a \sin\left(\frac{\pi}{2}(1-u)^3\right) \right] \qquad 0 \leqslant u \leqslant 1$$

$$(5.10)$$

$$Q_8(u) = \left[ a + b - b \cos\left(\frac{\pi}{2}u^3\right), \ -b \sin\left(\frac{\pi}{2}u^3\right) \right] \qquad 0 \leqslant u \leqslant 1 .$$

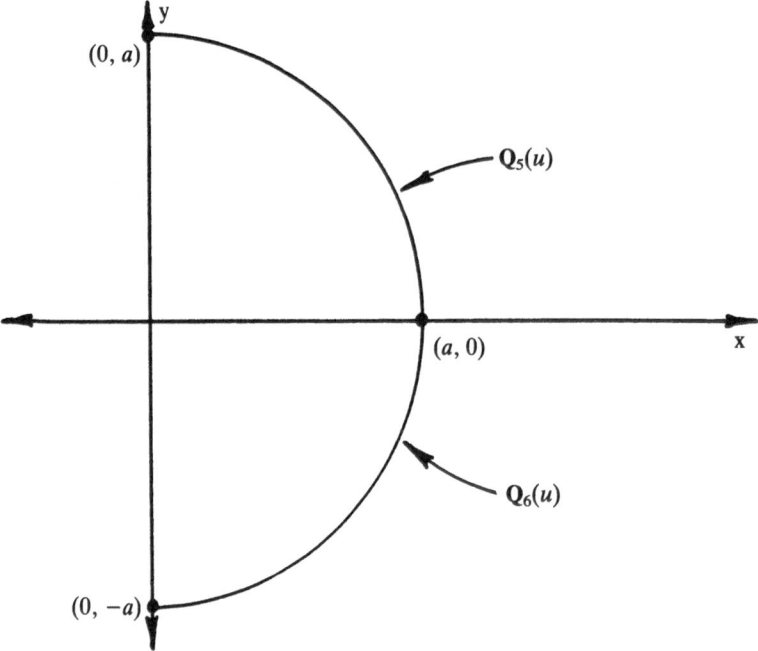

**Fig. 5.3.**  Discontinuous second derivative yet continuous curvature.

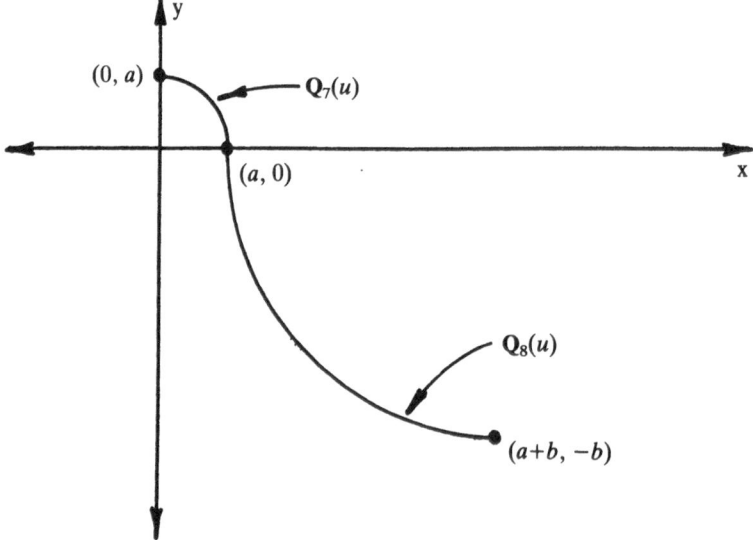

**Fig. 5.4.**  Continuous second derivative yet discontinuous curvature.

The parametric second derivative vectors are

$$\mathbf{Q}_7^{(2)}(u) = \left[ \frac{-9\pi^2 a(1-u)^4}{4} \cos\left(\frac{\pi}{2}(1-u)^3\right) \right.$$

$$- 3\pi a(1-u) \sin\left(\frac{\pi}{2}(1-u)^3\right),$$

$$\frac{-9\pi^2 a(1-u)^4}{4} \sin\left(\frac{\pi}{2}(1-u)^3\right)$$

$$\left. + 3\pi a(1-u) \cos\left(\frac{\pi}{2}(1-u)^3\right) \right] \qquad 0 \leqslant u \leqslant 1 \qquad (5.11)$$

$$\mathbf{Q}_8^{(2)}(u) = \left[ \frac{9\pi^2 b u^4}{4} \cos\left(\frac{\pi}{2}u^3\right) + 3\pi b u \sin\left(\frac{\pi}{2}u^3\right), \right.$$

$$\left. \frac{9\pi^2 b u^4}{4} \sin\left(\frac{\pi}{2}u^3\right) - 3\pi b u \cos\left(\frac{\pi}{2}u^3\right) \right] \qquad 0 \leqslant u \leqslant 1 \; .$$

Evaluating equation (5.11) at the joint yields

$$\mathbf{Q}_7^{(2)}(1) = [0, \; 0]$$

$$\mathbf{Q}_8^{(2)}(0) = [0, \; 0] \; . \qquad (5.12)$$

Hence the segments join with a continuous parametric second derivative vector even though the curvature vector is discontinuous since it jumps from $[-1/a, \; 0]$ for $\mathbf{Q}_7(u)$ to $[1/b, \; 0]$ for $\mathbf{Q}_8(u)$.

# 6 Geometric Continuity and Shape Parameters

## 6.1 Unit Tangent Vector

Given the two curves $\mathbf{Q}_1(u)$ and $\mathbf{Q}_2(u)$, consider the joint $\dfrac{\mathbf{Q}_1(u)}{\mathbf{Q}_2(u)}$. Recalling equation (4.1), continuity of the unit tangent vector is achieved if

$$\frac{\mathbf{Q}_1^{(1)}(1)}{|\mathbf{Q}_1^{(1)}(1)|} = \frac{\mathbf{Q}_2^{(1)}(0)}{|\mathbf{Q}_2^{(1)}(0)|} \tag{6.1}$$

that is,

$$\mathbf{Q}_1^{(1)}(1)\frac{|\mathbf{Q}_2^{(1)}(0)|}{|\mathbf{Q}_1^{(1)}(1)|} = \mathbf{Q}_2^{(1)}(0) \tag{6.2}$$

or

$$\mathbf{Q}_2^{(1)}(0) = \beta1\,\mathbf{Q}_1^{(1)}(1) \ . \tag{6.3}$$

## 6.2 Curvature Vector

Recall the expression for the curvature vector given in equation (4.36), and the constraints on the first derivative vector given in equation (6.3). The problem now is to determine the appropriate condition on $\mathbf{Q}_2^{(2)}(0)$ to maintain continuity of the curvature vector; that is,

$$\frac{(\beta1\mathbf{Q}_1^{(1)}(1) \times \mathbf{Q}_2^{(2)}(0)) \times \beta1\mathbf{Q}_1^{(1)}(1)}{|\beta1\mathbf{Q}_1^{(1)}(1)|^4}$$

$$= \frac{(\mathbf{Q}_1^{(1)}(1) \times \mathbf{Q}_1^{(2)}(1)) \times \mathbf{Q}_1^{(1)}(1)}{|\mathbf{Q}_1^{(1)}(1)|^4} \ . \tag{6.4}$$

It is easy to see that a sufficient solution is

$$\mathbf{Q}_2^{(2)}(0) = \beta1^2\mathbf{Q}_1^{(2)}(1) \ . \tag{6.5}$$

More generally, $\mathbf{Q}_2^{(2)}(0)$ can also include an additional term which is a multiple of

$Q_1^{(1)}(1)$:

$$Q_2^{(2)}(0) = \beta 1^2 Q_1^{(2)}(1) + \beta 2 Q_1^{(1)}(1) . \tag{6.6}$$

This condition guarantees continuity of the curvature vector (and hence of curvature).

Therefore the equations representing the constraints of the continuous unit tangent vector and curvature vector are

$$Q_2^{(1)}(0) = \beta 1 Q_1^{(1)}(1)$$
$$Q_2^{(2)}(0) = \beta 1^2 Q_1^{(2)}(1) + \beta 2 Q_1^{(1)}(1) . \tag{6.7}$$

These requirements can then replace the conventional conditions of continuous parametric first and second derivative vectors, respectively. Observe that these new constraints are expressed in terms of two inherent *shape parameters*, $\beta 1$ and $\beta 2$, in such a manner that $\beta 1 = 1$ indicates continuity of the parametric first derivative vector and $\beta 1 = 1$ and $\beta 2 = 0$ indicates continuity of the parametric first and second derivative vectors.

# 7 Derivation of the Beta-spline Curve Representation

## 7.1 Control Polygon and Control Graph

Like a B-spline [2, 20], a Beta-spline curve or surface is specified by a set of points called *control vertices*. Although these vertices do not generally lie on the generated curve or surface, their positions completely determine its shape. The vertices for a curve are an ordered sequence and are connected in succession to form a (closed or open) *control polygon* (Figure 7.1). This sequence of vertices will be denoted

$$V = [V_0, V_1, \ldots, V_m] \ .$$

A Beta-spline surface is defined by, but does not interpolate, a set of control vertices, in three-dimensional x-y-z space, which are organized as a two-dimensional graph with a rectangular topology. Each vertex is either an *interior* vertex or a *boundary* vertex. An interior vertex has four well-defined neighboring vertices. A boundary vertex has three neighboring vertices, except for the four corner vertices, each of which has only two neighbors.

This notion can be formalized quite elegantly by drawing on graph theory. The set of control vertices can be considered as a graph $\{V, E\}$ whose vertices form the set

$$V = \{V_{ij} | i = 0, 1, \ldots, m; \ j = 0, 1, \ldots, n\}$$

and with the set of edges

$$E = \{(V_{ij}, V_{i,j+1}) | i = 0, 1, \ldots, m; \ j = 0, 1, \ldots, n - 1\}$$
$$\cup \{(V_{ij}, V_{i+1,j}) | i = 0, 1, \ldots, m - 1; \ j = 0, 1, \ldots, n\} \ .$$

The interior vertices are the vertices

$$V_{ij} \quad \text{where} \quad 1 \leqslant i \leqslant m - 1 \quad \text{and} \quad 1 \leqslant j \leqslant n - 1$$

and the boundary vertices are

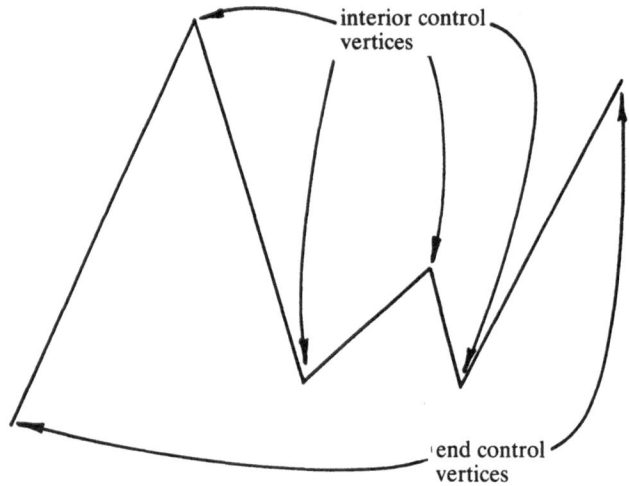

**Fig. 7.1.** A Beta-spline control polygon.

$\mathbf{V}_{0j}, \quad j = 0, 1, \ldots, n - 1;$

$\mathbf{V}_{in}, \quad i = 0, 1, \ldots, m - 1;$

$\mathbf{V}_{mj}, \quad j = 1, 2, \ldots, n; \quad \text{and}$

$\mathbf{V}_{i0}, \quad i = 1, 2, \ldots, m \ .$

To emphasize this graph theoretic interpretation, the author has chosen the term
*control graph* to describe the set of control vertices (Figure 7.2).

The generated curve or surface tends to mimic the overall shape of the control
polygon or graph, and the manipulation of a control vertex causes a predictable
modification in the resulting shape.

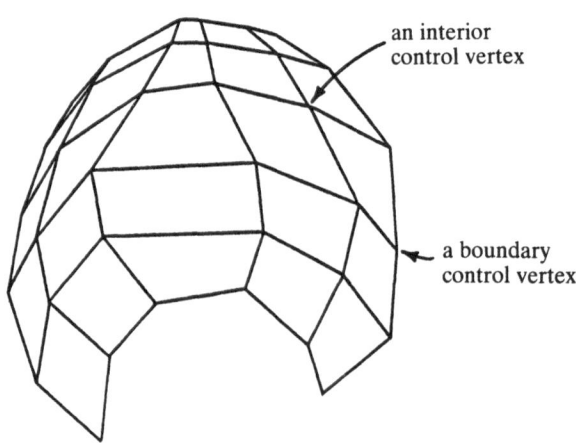

**Fig. 7.2.** A Beta-spline control graph.

## 7.2  Local Control

The Beta-spline basis is a local basis; that is, each Beta-spline basis function has *local support* (nonzero over a minimal number of spans). Since each control vertex is associated with a basis function, it only influences a local portion of the curve or surface and has no effect on the remainder of it. The effect of moving a single control vertex, then, is localized to a predetermined portion of the curve or surface. This enables the user of the Beta-spline representation to have precise control over the resulting shape by moving one control vertex at a time. This action consequently modifies only a local portion without the undesired side effect of disturbing the other portions. Moreover, since only part of the curve or surface is affected, only that part need be recomputed. This is much more computationally efficient than what is required to modify global formulations where any change necessitates the recomputation of the entire curve or surface.

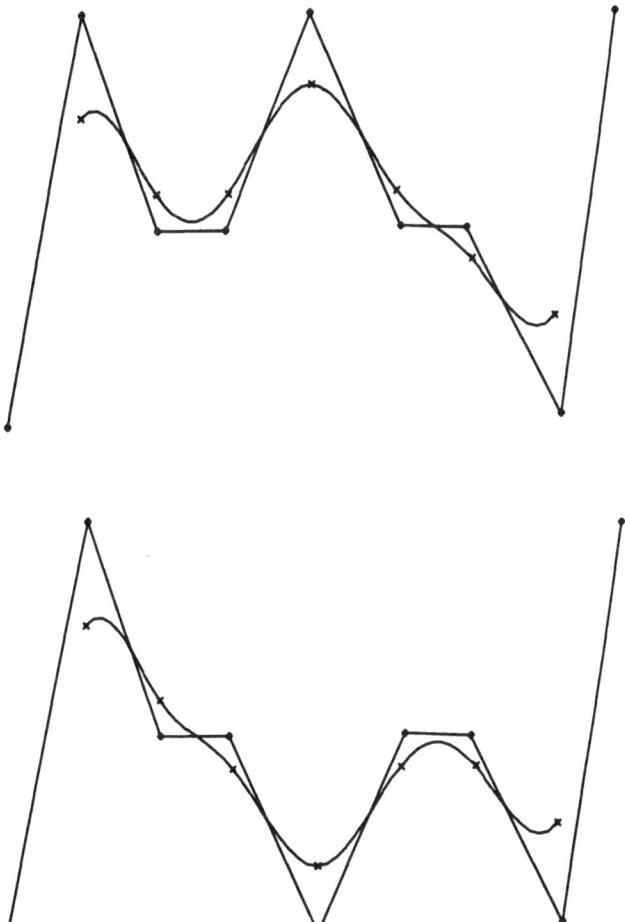

**Fig. 7.3.**   The effects of moving a Beta-spline control vertex are confined to four segments.

The Beta-spline formulation exploits the piecewise representation, in order to achieve local control, by defining each piece in terms of only a few nearby vertices. For Beta-spline curves, each curve segment is controlled by only four of the control vertices and is completely unaffected by all the other control vertices. Equivalently, a given control vertex only influences four curve segments and has no effect whatsoever on the remaining segments. This means that the effects of moving a Beta-spline control vertex are confined to four segments, as shown in Figure 7.3.

A Beta-spline surface has each surface patch controlled by sixteen control vertices and is unaffected by all other control vertices. Again, this is equivalent to the fact that a given control vertex exerts influence over only sixteen surface patches and has no effect on the remaining patches. Thus, the effects of manipulating one control vertex are limited to sixteen patches.

## 7.3 Explanation

In order to effect local control, a Beta-spline curve segment is completely controlled by only four of the control vertices; therefore, a point on this curve segment can be regarded as a weighted average of these four control vertices. Associated with each control vertex is a weighting factor which is a scalar-valued function evaluated at some value of the *domain* parameter $u$. For a Beta-spline curve segment, this parameter indicates the location on the segment as it varies from a value of zero at the beginning of the segment to a value of unity at the end. The $i^{th}$ Beta-spline curve segment will be denoted $Q_i(u)$.

The geometric continuity constraints derived in Chapter 6 can be applied to the joint of the $i^{th}$ and $(i + 1)^{st}$ curve segments yielding the following conditions:

$$Q_{i+1}(0) = Q_i(1)$$

$$Q_{i+1}^{(1)}(0) = \beta 1 Q_i^{(1)}(1) \tag{7.1}$$

$$Q_{i+1}^{(2)}(0) = \beta 1^2 Q_i^{(2)}(1) + \beta 2 Q_i^{(1)}(1) .$$

$\beta 1$ and $\beta 2$ will be called *shape parameters* because they control the shape of the curve. Since these constraints (equation (7.1)) are expressed in terms of the given shape parameters, the latter must appear in the expressions for the Beta-spline curve segment $Q_i(u)$.

Let the control polygon be composed of the sequence of control vertices $[V_0, V_1, \ldots, V_m]$. Then a point on the $i^{th}$ curve segment is a weighted average of the four control vertices $V_{i+r}$, $r = -2, -1, 0, 1$. The coordinates of the point $Q_i(u)$ on the $i^{th}$ curve segment are then given by

$$Q_i(u) = \sum_{r=-2}^{1} b_r(\beta 1, \beta 2; u) V_{i+r} \qquad \text{for } 0 \leqslant u < 1 . \tag{7.2}$$

As the domain parameter $u$ varies from zero to unity, the $i^{th}$ curve segment is traced out. (Figure 7.4 shows the $(i - 1)^{st}$, $i^{th}$, and $(i + 1)^{st}$ Beta-spline curve segments.)

The weighting factors are the scalar-valued *basis functions* $b_r(\beta 1, \beta 2; u)$, $r = -2, -1, 0, 1$, evaluated at some value of the domain parameter $u$, and of each shape parameter $\beta 1$ and $\beta 2$. If $\beta 1 > 0$ and $\beta 2 \geqslant 0$, they form a basis; that is, they are linearly independent, and any possible Beta-spline curve segment can be expressed

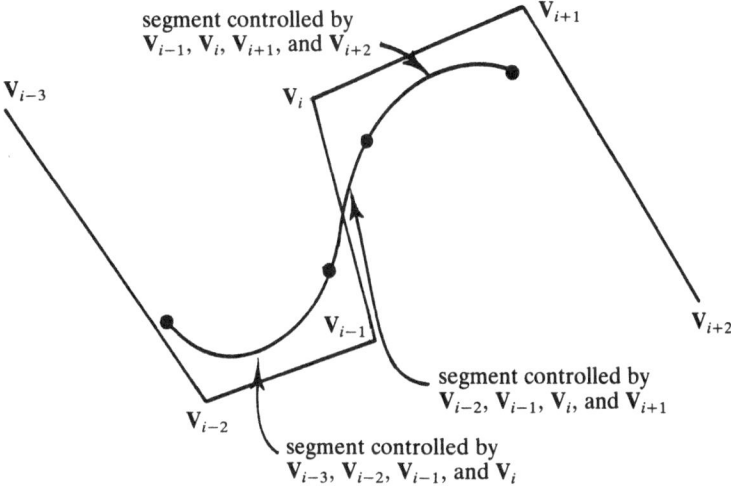

**Fig. 7.4.** Beta-spline curve with its control polygon showing the control vertices controlling each segment.

as a linear combination of them. Moreover, the combination coefficients of this linear combination are unique since the basis functions are linearly independent. Therefore, every Beta-spline curve segment with $\beta1 > 0$ and $\beta2 \geqslant 0$ has a unique representation as a linear combination of these basis functions, where the combination coefficients are the associated control vertices. The Beta-spline basis functions are derived in the following section.

The Beta-spline curve segment $Q_i(u)$ is controlled by the control vertices $V_{i+r}, r = -2, -1, 0, 1$. For the next curve segment, $Q_{i+1}(u)$, the first of these control vertices, $V_{i-2}$, is dropped, and a new control vertex, $V_{i+2}$, is added. The basis functions are shifted to this new set of control vertices so that the basis functions, $b_r(\beta1, \beta2; u), r = -2, -1, 0, 1$, are now associated with the control vertices $V_{i+r}, r = -1, 0, 1, 2$, respectively.

## 7.4 Derivation of the Beta-spline Basis Functions

Each basis function is a function of $\beta1$ and $\beta2$ and of $u$ such that it is a cubic polynomial in $u$ whose polynomial coefficients are themselves functions of $\beta1$ and $\beta2$:

$$b_r(\beta1, \beta2; u) = \sum_{g=0}^{3} c_{gr}(\beta1, \beta2)u^g \quad \text{for } 0 \leqslant u < 1 \quad \text{for } r = -2, -1, 0, 1 \ . \quad (7.3)$$

The coefficients $c_{gr}(\beta1, \beta2), g = 0, 1, 2, 3$ and $r = -2, -1, 0, 1$, are functions of $\beta1$ and $\beta2$. An explicit expression in terms of these shape parameters must be determined for each one. This can be done so as to satisfy the geometric continuity constraints.

From equation (7.2), the mathematical formulations for a Beta-spline segment $Q_i(u)$ and its first and second derivative vectors, $Q_i^{(1)}(u)$ and $Q_i^{(2)}(u)$, are given by

$$Q_i^{(a)}(u) = \sum_{r=-2}^{1} b_r^{(a)}(\beta1, \beta2; u)V_{i+r} \qquad \text{for } 0 \leqslant u < 1 \quad \text{for } a = 0, 1, 2 . \quad (7.4)$$

Evaluating equation (7.4) at $u = 0$ and $u = 1$ for $a = 0, 1, 2$ and substituting it into equation (7.1) yields

$$\sum_{r=-2}^{1} b_r(\beta1, \beta2; 0)V_{i+1+r} = \sum_{r=-2}^{1} b_r(\beta1, \beta2; 1)V_{i+r}$$

$$\sum_{r=-2}^{1} b_r^{(1)}(\beta1, \beta2; 0)V_{i+1+r} = \beta1 \sum_{r=-2}^{1} b_r^{(1)}(\beta1, \beta2; 1)V_{i+r}$$

$$\sum_{r=-2}^{1} b_r^{(2)}(\beta1, \beta2; 0)V_{i+1+r} = \beta1^2 \sum_{r=-2}^{1} b_r^{(2)}(\beta1, \beta2; 1)V_{i+r}$$

$$+ \beta2 \sum_{r=-2}^{1} b_r^{(1)}(\beta1, \beta2; 1)V_{i+r} .$$

(7.5)

Equating coefficients of the vertices $V_{i+r}$, $r = -2, -1, 0, 1, 2$, results in

$$0 = b_{-2}(\beta1, \beta2; 1)$$

$$b_{r-1}(\beta1, \beta2; 0) = b_r(\beta1, \beta2; 1) \qquad\qquad\qquad \text{for } r = -1, 0, 1$$

$$b_1(\beta1, \beta2; 0) = 0$$

$$0 = \beta1 b_{-2}^{(1)}(\beta1, \beta2; 1)$$

$$b_{r-1}^{(1)}(\beta1, \beta2; 0) = \beta1 b_r^{(1)}(\beta1, \beta2; 1) \qquad\qquad \text{for } r = -1, 0, 1$$

$$b_1^{(1)}(\beta1, \beta2; 0) = 0$$

$$0 = \beta1^2 b_{-2}^{(2)}(\beta1, \beta2; 1) + \beta2 b_{-2}^{(1)}(\beta1, \beta2; 1)$$

$$b_{r-1}^{(2)}(\beta1, \beta2; 0) = \beta1^2 b_r^{(2)}(\beta1, \beta2; 1)$$

$$+ \beta2 b_r^{(1)}(\beta1, \beta2; 1) \qquad\qquad\qquad \text{for } r = -1, 0, 1$$

$$b_1^{(2)}(\beta1, \beta2; 0) = 0 .$$

(7.6)

Performing the necessary differentiations of the basis functions, and evaluating at $u = 0$ and $u = 1$ yields the following system of equations:

$$c_{3,-2}(\beta1, \beta2) + c_{2,-2}(\beta1, \beta2) + c_{1,-2}(\beta1, \beta2) + c_{0,-2}(\beta1, \beta2) = 0$$

$$c_{0,r-1}(\beta1, \beta2) = c_{3,r}(\beta1, \beta2) + c_{2,r}(\beta1, \beta2) + c_{1,r}(\beta1, \beta2)$$

$$+ c_{0,r}(\beta1, \beta2) \qquad\qquad\qquad \text{for } r = -1, 0, 1$$

$$c_{0,1}(\beta1, \beta2) = 0$$

$$\beta1(3c_{3,-2}(\beta1, \beta2) + 2c_{2,-2}(\beta1, \beta2) + c_{1,-2}(\beta1, \beta2)) = 0$$

$$c_{1,r-1}(\beta1, \beta2) = \beta1(3c_{3,r}(\beta1, \beta2) + 2c_{2,r}(\beta1, \beta2)$$

$$+ c_{1,r}(\beta1, \beta2)) \qquad\qquad\qquad \text{for } r = -1, 0, 1$$

(7.7)

$$c_{1,1}(\beta 1, \beta 2) = 0$$

$$3(2\beta 1^2 + \beta 2)c_{3,-2}(\beta 1, \beta 2) + 2(\beta 1^2 + \beta 2)c_{2,-2}(\beta 1, \beta 2)$$
$$+ \beta 2 c_{1,-2}(\beta 1, \beta 2) = 0$$

$$2c_{2,r-1}(\beta 1, \beta 2) = 3(2 \beta 1^2 + \beta 2)c_{3,r}(\beta 1, \beta 2) + 2(\beta 1^2 + \beta 2)c_{2,r}(\beta 1, \beta 2)$$
$$+ \beta 2 c_{1,r}(\beta 1, \beta 2) \qquad\qquad\qquad \text{for } r = -1, 0, 1$$

$$c_{2,1}(\beta 1, \beta 2) = 0 \ .$$

This is a system of fifteen linear equations in terms of the sixteen unknown coefficient functions $c_{gr}(\beta 1, \beta 2)$, $g = 0, 1, 2, 3$ and $r = -2, -1, 0, 1$, and the *given* shape parameters $\beta 1$ and $\beta 2$. One more constraint is required in order to uniquely determine the coefficient functions. A useful constraint for axis independence and convex hull properties is to normalize the basis functions (that is, require that they sum to unity), at $u = 0$:

$$c_{0,-2}(\beta 1, \beta 2) + c_{0,-1}(\beta 1, \beta 2) + c_{0,0}(\beta 1, \beta 2) + c_{0,1}(\beta 1, \beta 2) = 1 \ . \qquad (7.8)$$

Noting that three of these equations in (7.7) identically set a coefficient function to zero, these equations and coefficient functions can be eliminated from the system, thereby reducing it to a set of thirteen equations in thirteen unknown coefficient functions. These equations are expressed in terms of the given shape parameters which will appear in the expressions for the basis functions. Thus, the determination of these coefficient functions requires symbolic, not numeric, computation. To avoid pages of tedious (and probably incorrect) algebra, REDUCE [12, 13, 14, 25], a computer algebra system developed at the Department of Computer Science at the University of Utah, was employed. The main REDUCE program is included in the Appendix; using this approach, the thirteen unknown coefficient functions are determined to be:

$$c_{0,-2}(\beta 1, \beta 2) = \frac{2 \beta 1^3}{\delta}$$

$$c_{1,-2}(\beta 1, \beta 2) = -\frac{6 \beta 1^3}{\delta}$$

$$c_{2,-2}(\beta 1, \beta 2) = \frac{6 \beta 1^3}{\delta}$$

$$c_{3,-2}(\beta 1, \beta 2) = -\frac{2 \beta 1^3}{\delta}$$

$$c_{0,-1}(\beta 1, \beta 2) = \frac{4 \beta 1^2 + 4 \beta 1 + \beta 2}{\delta}$$

$$c_{1,-1}(\beta 1, \beta 2) = \frac{6 \beta 1(\beta 1^2 - 1)}{\delta}$$

$$c_{2,-1}(\beta 1, \beta 2) = 3\frac{-2\,\beta 1^3 - 2\,\beta 1^2 - \beta 2}{\delta} \tag{7.9}$$

$$c_{3,-1}(\beta 1, \beta 2) = 2\frac{\beta 1^3 + \beta 1^2 + \beta 1 + \beta 2}{\delta}$$

$$c_{0,0}(\beta 1, \beta 2) = \frac{2}{\delta}$$

$$c_{1,0}(\beta 1, \beta 2) = \frac{6\,\beta 1}{\delta}$$

$$c_{2,0}(\beta 1, \beta 2) = 3\frac{2\,\beta 1^2 + \beta 2}{\delta}$$

$$c_{3,0}(\beta 1, \beta 2) = -2\frac{\beta 1^2 + \beta 1 + \beta 2 + 1}{\delta}$$

$$c_{3,1}(\beta 1, \beta 2) = \frac{2}{\delta}$$

where

$$\delta = 2\,\beta 1^3 + 4\,\beta 1^2 + 4\,\beta 1 + \beta 2 + 2 \ .$$

An efficient algorithm is given below that computes the four Beta-spline basis functions for a given value of each shape parameter, $\beta 1$ and $\beta 2$, and of the domain parameter $u$. It requires 16 multiplications, 23 additions/subtractions, and 8 divisions to evaluate $\delta$ and the four basis functions at a given value of each shape parameter and domain parameter. For easier reference, this algorithm has been separated into the following sequence of procedure calls:

     compute_d (beta1, beta2, d);

     compute_delta (d, delta);

     compute_c (delta, d, c);

     compute_b (c, u, b);

where the procedures are defined as follows:

     **procedure** compute_d (beta1; beta2; d);

     **begin** (∗ compute_d ∗)

        $t_1 := beta1^2 + beta2;$

        $t_2 := beta1^2 + t_1;$

        $d_{0,-2} := beta1 ∗ (t_2 - beta2);$

        $d_{0,0} := 2;$

        $d_{1,0} := 6 ∗ beta1;$

        $d_{2,0} := 3 ∗ t_2;$

        $d_{3,0} := -2 ∗ (t_1 + beta1 + 1);$

$$d_{0,-1} := d_{1,0} + d_{2,0} + d_{3,0} + 2;$$
$$t_3 := 3 * d_{0,-2};$$
$$d_{1,-1} := t_3 - d_{1,0};$$
$$d_{2,-1} := -t_3 - d_{2,0};$$
$$d_{3,-1} := d_{0,-2} - d_{3,0} - 2$$

**end** (* compute_d *);

and

**procedure** compute_delta (d; *delta*);
**begin** (* compute_delta *)
$$delta := d_{0,-2} + d_{0,-1} + 2$$
**end** (* compute_delta *);

and

**procedure** compute_c (*delta; d; c*);
**begin** (* compute_c *)
  **for** $g := 0$ **to** 3 **do**
  **begin**
    $$c_{g,0} := d_{g,0}/delta;$$
    $$c_{g,-1} := c_{g,-1}/delta$$
  **end**;
  $$c_{3,1} := c_{0,0}$$
**end** (* compute_c *);

and

**procedure** compute_b (*c; u; b*);
**begin** (* compute_b *)
  $$b_1 := c_{3,1} * u^3;$$
  $$b_0 := c_{0,0} + u * (c_{1,0} + u * (c_{2,0} + u * c_{3,0}));$$
  $$b_{-1} := c_{0,-1} + u * (c_{1,-1} + u * (c_{2,-1} + u * c_{3,-1}));$$
  $$b_{-2} := 1 - b_1 - b_0 - b_{-1}$$
**end** (* compute_b *);

These four algorithms require 7 multiplications and 12 additions/subtractions; 2 additions/subtractions; 8 divisions; and 9 multiplications and 9 additions/subtractions, respectively.

Explicit expressions for the Beta-spline basis functions can be written as follows:

$$b_{-2}(\beta 1, \beta 2; u) = \frac{2\,\beta 1^3(1-u)^3}{\delta}$$

$$b_{-1}(\beta 1, \beta 2; u) = \frac{1}{\delta}[2 \beta 1^3 u(u^2 - 3u + 3) + 2 \beta 1^2(u^3 - 3u^2 + 2)$$

$$+ 2 \beta 1(u^3 - 3u + 2) + \beta 2(2u^3 - 3u^2 + 1)]$$

$$b_0(\beta 1, \beta 2; u) = \frac{1}{\delta}[2 \beta 1^2 u^2(-u + 3) + 2 \beta 1 u(-u^2 + 3)$$

$$+ \beta 2 u^2(-2u + 3) + 2(-u^3 + 1)]$$

$$b_1(\beta 1, \beta 2; u) = \frac{2u^3}{\delta} .$$

(7.10)

Evaluating at the extreme values $u = 0$ and $u = 1$ of the domain parameter yields:

$$b_{-2}(\beta 1, \beta 2; 0) = \frac{2 \beta 1^3}{\delta}$$

$$b_{-1}(\beta 1, \beta 2; 0) = \frac{4 \beta 1^2 + 4 \beta 1 + \beta 2}{\delta}$$

$$b_0(\beta 1, \beta 2; 0) = \frac{2}{\delta}$$

$$b_1(\beta 1, \beta 2; 0) = 0$$

$$b_{-2}(\beta 1, \beta 2; 1) = 0$$

(7.11)

$$b_{-1}(\beta 1, \beta 2; 1) = \frac{2 \beta 1^3}{\delta}$$

$$b_0(\beta 1, \beta 2; 1) = \frac{4 \beta 1^2 + 4 \beta 1 + \beta 2}{\delta}$$

$$b_1(\beta 1, \beta 2; 1) = \frac{2}{\delta} .$$

The first derivative of each Beta-spline basis function is as follows:

$$b_{-2}^{(1)}(\beta 1, \beta 2; u) = -\frac{6 \beta 1^3(1 - u)^2}{\delta}$$

$$b_{-1}^{(1)}(\beta 1, \beta 2; u) = \frac{6}{\delta}[\beta 1^3(u^2 - 2u + 1) + \beta 1^2 u(u - 2)$$

$$+ \beta 1(u^2 - 1) + \beta 2 u(u - 1)]$$

$$b_0^{(1)}(\beta 1, \beta 2; u) = \frac{6}{\delta}[\beta 1^2 u(-u + 2) + \beta 1(-u^2 + 1)$$

$$+ \beta 2 u(-u + 1) - u^2]$$

$$b_1^{(1)}(\beta 1, \beta 2; u) = \frac{6u^2}{\delta} .$$

(7.12)

Evaluating at the extreme values $u = 0$ and $u = 1$ of the domain parameter yields:

$$b_{-2}^{(1)}(\beta 1, \beta 2; 0) = -\frac{6\,\beta 1^3}{\delta}$$

$$b_{-1}^{(1)}(\beta 1, \beta 2; 0) = \frac{6\,\beta 1(\beta 1^2 - 1)}{\delta}$$

$$b_0^{(1)}(\beta 1, \beta 2; 0) = \frac{6\,\beta 1}{\delta}$$

$$b_1^{(1)}(\beta 1, \beta 2; 0) = 0$$

$$b_{-2}^{(1)}(\beta 1, \beta 2; 1) = 0$$

$$b_{-1}^{(1)}(\beta 1, \beta 2; 1) = -\frac{6\,\beta 1^2}{\delta}$$

$$b_0^{(1)}(\beta 1, \beta 2; 1) = \frac{6(\beta 1^2 - 1)}{\delta}$$

$$b_1^{(1)}(\beta 1, \beta 2; 1) = \frac{6}{\delta}\ .$$

(7.13)

The second derivative of each Beta-spline basis function is as follows:

$$b_{-2}^{(2)}(\beta 1, \beta 2; u) = \frac{12\,\beta 1^3(1 - u)}{\delta}$$

$$b_{-1}^{(2)}(\beta 1, \beta 2; u) = \frac{6}{\delta}[2\,\beta 1^3(u - 1) + 2\,\beta 1^2(u - 1) + 2\,\beta 1 u + \beta 2(2u - 1)]$$

$$b_0^{(2)}(\beta 1, \beta 2; u) = \frac{6}{\delta}[2\,\beta 1^2(-u + 1) - 2\,\beta 1 u + \beta 2(-2u + 1) - 2u]$$

$$b_1^{(2)}(\beta 1, \beta 2; u) = \frac{12u}{\delta}\ .$$

(7.14)

Evaluating at the extreme values $u = 0$ and $u = 1$ of the domain parameter yields:

$$b_{-2}^{(2)}(\beta 1, \beta 2; 0) = \frac{12\,\beta 1^3}{\delta}$$

$$b_{-1}^{(2)}(\beta 1, \beta 2; 0) = \frac{6}{\delta}[-2\,\beta 1^3 - 2\,\beta 1^2 - \beta 2]$$

$$b_0^{(2)}(\beta 1, \beta 2; 0) = \frac{6}{\delta}[2\,\beta 1^2 + \beta 2]$$

$$b_1^{(2)}(\beta 1, \beta 2; 0) = 0$$

$$b_{-2}^{(2)}(\beta 1, \beta 2; 1) = 0$$

(7.15)

$$b_{-1}^{(2)}(\beta1, \beta2; 1) = \frac{6}{\delta}[2\,\beta1 + \beta2]$$

$$b_0^{(2)}(\beta1, \beta2; 1) = \frac{6}{\delta}[-2\,\beta1 - \beta2 - 2]$$

$$b_2^{(2)}(\beta1, \beta2; 1) = \frac{12}{\delta} \ .$$

These expressions at the extreme values will be needed for the analysis of end conditions in Chapter 11 and of boundary conditions in Chapter 16.

## 7.5  Convex Hull Property

Recall that the last constraint to determine the coefficient functions was that the basis functions should sum to unity at $u = 0$ (equation (7.8)). However, an even stronger condition has been satisfied by the basis functions; they sum to unity for *all* values of the domain parameter $u$,

$$\sum_{r=-2}^{1} b_r(\beta1, \beta2; u) = 1 \ . \tag{7.16}$$

Specifically, for each nonzero power of $u$ in the basis functions, the coefficient functions (equation (7.9)) sum to zero,

$$\sum_{r=-2}^{1} c_{gr}(\beta1, \beta2) = 0 \qquad\qquad \text{for } g = 1, 2, 3 \ , \tag{7.17}$$

and the coefficient functions of the constant terms add to unity,

$$\sum_{r=-2}^{1} c_{0r}(\beta1, \beta2) = 1 \ . \tag{7.18}$$

Now, consider the basis functions written in the form given in equation (7.10). All the expressions in $u$ are nonnegative for $0 \leqslant u < 1$. Therefore, each of the basis functions is nonnegative for $0 \leqslant u < 1$ as long as $\beta1 \geqslant 0$ and $\beta2 \geqslant 0$. Since each basis function is nonnegative for these values, and since they sum to unity, each one

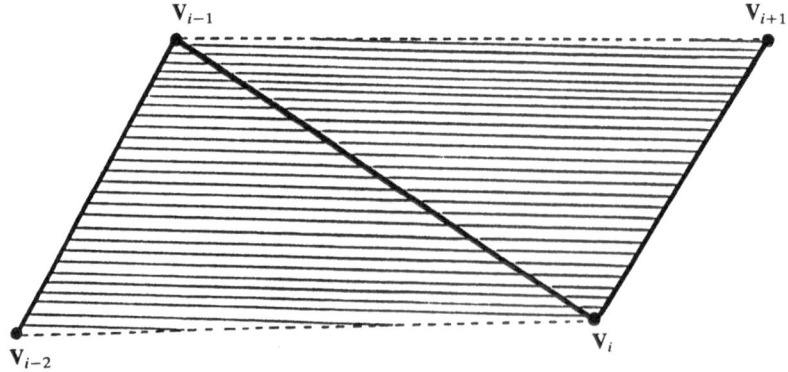

**Fig. 7.5.**  Convex hull of four control vertices.

can only attain values between zero and unity. These results are sufficient to prove that a point on a Beta-spline curve segment is a convex combination of control vertices; specifically, the $i^{th}$ segment is a convex combination of the four control vertices $\mathbf{V}_{i+r}, r = -2, -1, 0, 1$.

The set of all possible convex combinations of these vertices is their convex hull; that is, the "smallest" convex set which contains these vertices. Thus, each Beta-spline curve segment lies within the convex hull of its four defining control vertices. Figure 7.5 shows the convex hull of the four control vertices $\mathbf{V}_{i+r}, r = -2, -1, 0, 1$, which will contain the $i^{th}$ Beta-spline segment.

Since a Beta-spline curve is composed of a sequence of such segments, it follows that the entire Beta-spline curve will be contained in the union of the convex hulls of each successive set of four control vertices (Figure 7.6).

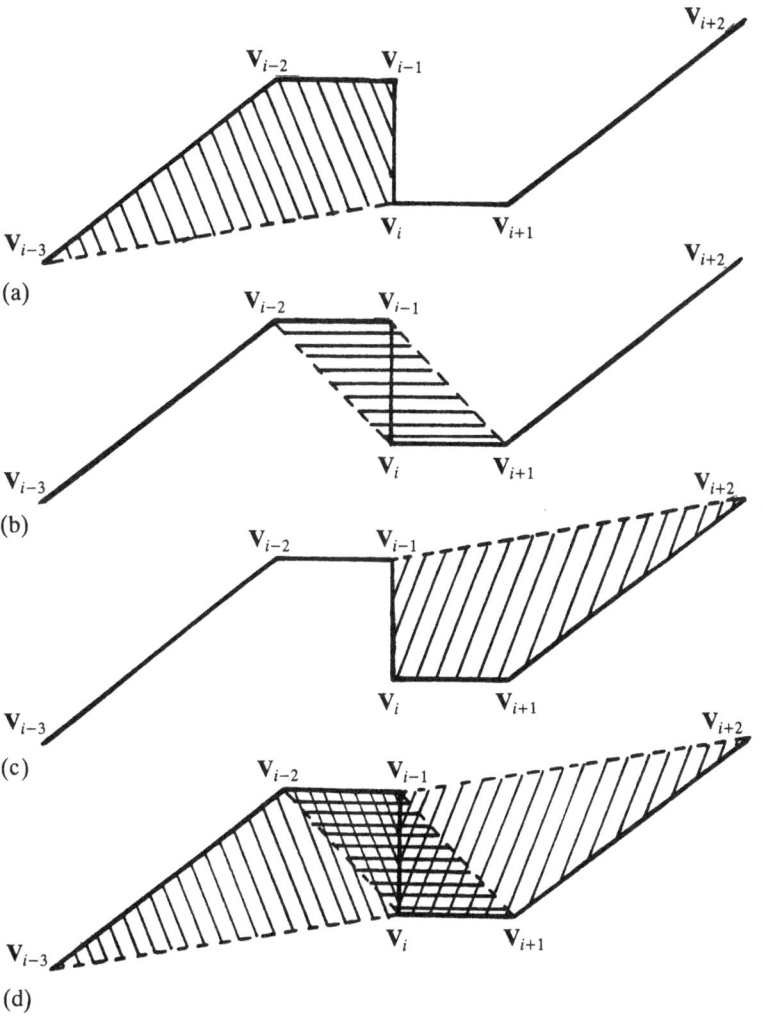

**Fig. 7.6.**   Union of convex hulls of each of three successive sets of four control vertices.

# 8 Curve Evaluation and Perturbation with Uniform Shape Parameters

## 8.1 Evaluation Method I

$\beta 1$ and $\beta 2$ have been assumed to be *uniform* shape parameters, each assuming a single value over the entire curve. This assumption can be exploited to efficiently evaluate a Beta-spline curve. Observe that all the coefficient functions have a constant denominator of $\delta$. Thus, all the divisions can be performed prior to the actual computation of the Beta-spline basis functions. The following algorithm evaluates the basis functions at $p + 1$ given values of the domain parameter $u$, for a given value of each uniform shape parameter, $\beta 1$ and $\beta 2$, and requires $7 + 9(p + 1)$ multiplications, $12 + 2 + 9(p + 1)$ additions/subtractions, and 8 divisions.

    compute_d (*beta1, beta2, d*);

    compute_delta (*d, delta*);

    compute_c (*delta, d, c*);

    **for each** $u$ **in** $\{u_0, u_1, \ldots, u_p\}$ **do** compute_b (*c, u, b*);

where these algorithms were given in Section 7.4.

Now the basis functions have been evaluated for a set of $p + 1$ given values of the domain parameter $u$. The values of the functions at each of these parametric values can then be used to compute a corresponding set of points on a single curve segment. Equation (7.1) can be used in the design of the following algorithm which requires $4d$ multiplications and $3d$ additions/subtractions to compute a curve point at the domain parametric value $u$, where $d$ is the dimension of the curve.

    **procedure** compute_Q ($i$; $b$; $V$; $Q$);

    **begin** ($*$ compute_Q $*$)

        $Q_i := b_{-2} * V_{i-2}$;

        **for** $r := -1$ **to** 1 **do** $Q_i := Q_i + b_r * V_{i+r}$

    **end** ($*$ compute_Q $*$);

Now consider the following algorithm which computes a general curve composed of $m$ segments where the $i^{th}$ segment is evaluated at $p_i + 1$ values of the domain parameter:

compute_d (*beta*1, *beta*2, *d*);
compute_delta (*d*, *delta*);
compute_c (*delta*, *d*, *c*);
**for** $i := 1$ **to** $m$ **do**
    **for each** $u$ **in** $\{u_{i_k}|k = 0, 1, \ldots, p_i\}$ **do**
    **begin**
        compute_b (*c*, *u*, *b*);
        compute_Q (*i*, *b*, **V**, **Q**)
    **end**;

This complete algorithm requires a total of $7 + \sum_{i=1}^{m} (p_i + 1)(9 + 4d)$ multiplications, $12 + 2 + \sum_{i=1}^{m} (p_i + 1)(9 + 3d)$ additions/subtractions, and 8 divisions. For comparison purposes, the average *density* of evaluation, $p$, is defined as:

$$p = \frac{1}{m} \sum_{i=1}^{m} p_i . \tag{8.1}$$

Then these computational requirements become $7 + m(p + 1)(9 + 4d)$ multiplications, $14 + 3m(p + 1)(3 + d)$ additions/subtractions, and 8 divisions.

Observe that a more efficient algorithm would be possible if all curve segments were evaluated at the same set of values of the domain parameter. Note that this does not impose any restriction on the selection of a particular set of values of the domain parameter. The following algorithm requires $7 + (p + 1)(9 + 4dm)$ multiplications, $14 + 3(p + 1)(3 + dm)$ additions/subtractions, and 8 divisions to evaluate the curve at $p + 1$ values of the domain parameter on each of the $m$ segments.

compute_d (*beta*1, *beta*2, *d*);
compute_delta (*d*, *delta*);
compute_c (*delta*, *d*, *c*);
**for each** $u$ **in** $\{u_0, u_1, \ldots, u_p\}$ **do**
**begin**
    compute_b (*c*, *u*, *b*);
    **for** $i := 1$ **to** $m$ **do** compute_Q (*i*, *b*, **V**, **Q**)
**end**;

## 8.2  Evaluation Method II

Substituting equation (7.3) into equation (7.2) yields

$$Q_i(u) = \sum_{r=-2}^{1} \left( \sum_{g=0}^{3} c_{gr}(\beta 1, \beta 2)u^g \right) V_{i+r} . \tag{8.2}$$

Rearranging,

$$Q_i(u) = \sum_{g=0}^{3} \left( \sum_{r=-2}^{1} c_{gr}(\beta1, \beta2)V_{i+r} \right) u^g . \tag{8.3}$$

Denoting the inner summation by $S_{ig}(\beta1, \beta2)$,

$$S_{ig}(\beta1, \beta2) = \sum_{r=-2}^{1} c_{gr}(\beta1, \beta2)V_{i+r} \qquad\qquad g = 0, 1, 2, 3 . \tag{8.4}$$

Then

$$Q_i(u) = \sum_{g=0}^{3} S_{ig}(\beta1, \beta2)u^g . \tag{8.5}$$

Rewriting equation (8.5) to reduce the number of multiplications,

$$Q_i(u) = S_{i0}(\beta1, \beta2) + u(S_{i1}(\beta1, \beta2) + u(S_{i2}(\beta1, \beta2) + u(S_{i3}(\beta1, \beta2)))) . \tag{8.6}$$

Using these equations, the following algorithm computes a curve composed of $m$ segments where the $i^{\text{th}}$ segment is evaluated at $p_i + 1$ values of the domain parameter:

```
compute_d (beta1, beta2, d);
compute_delta (d, delta);
compute_all_c (delta, d, c);
for i := 1 to m do
begin
   for g := 0 to 3 do
   begin
      Sig := cg,-2 * Vi-2;
      for r := -1 to 1 do
         Sig := Sig + cgr * Vi+r
   end;
   for each u in {uik|k = 0, 1, ..., pi} do
   begin
      Qi := Si3;
      for g := 2 downto 0 do Qi := Sig + u * Qi
   end
end;
```

where procedure compute_all_c (delta; d; c) is defined as follows:

**procedure** compute_all_c (*delta*; *d*; *c*);

**begin** (∗ compute_all_c ∗)

    **for** $g := 0$ **to** 3 **do**

    **begin**

        $c_{g,0} := d_{g,0}/delta$;

        $c_{g,-1} := d_{g,-1}/delta$

    **end**;

    $c_{0,-2} := d_{0,-2}/delta$;

    $c_{3,1} := c_{0,0}$;

    $c_{3,-2} := -c_{0,-2}$;

    $c_{2,-2} := 3 * c_{3,-2}$;

    $c_{1,-2} := -c_{2,-2}$;

    $c_{2,1} := 0$;

    $c_{1,1} := 0$;

    $c_{0,1} := 0$

**end** (∗ compute_all_c ∗);

This algorithm requires $8 + \sum_{i=1}^{m} d\{4(1+3) + (p_i + 1)3\}$ multiplications, $12 + 2 + \sum_{i=1}^{m} d\{4(3) + (p_i + 1)3\}$ additions/subtractions, and 9 divisions. Using the notation defined in equation (8.1), and rearranging, these computational requirements become $8 + dm(16 + 3(p + 1))$ multiplications, $14 + 3dm(4 + (p + 1))$ additions/subtractions, and 9 divisions.

It should be noted that this algorithm computes the general curve evaluated at a different set of values of the domain parameter on each segment, with no loss of efficiency.

## 8.3  Comparison of Evaluation Methods I and II

As mentioned in Section 8.2, evaluation method II naturally computes the general curve evaluated at a different set of values of the domain parameter for each segment. In this general case, evaluation method II is more efficient than method I.

If all curve segments were evaluated at the same set of values of the domain parameter, the comparison is far more subtle. The computational requirements of evaluation method II are unchanged in this case. However, a more efficient algorithm for evaluation method I for this situation was designed in Section 8.1. A careful comparison of the computational requirements of these algorithms shows that neither algorithm is more efficient in general. Which algorithm is preferable is dependent upon the number of segments ($m$) and the number of curve points to be evaluated per segment ($p + 1$). Specifically, evaluation method I is more efficient if there are many segments or if evaluation is sparse (large $m/p$ ratio), and evaluation method II is preferable if there are only a few segments or if evaluation is dense (small $m/p$ ratio).

## 8.4 Perturbation Due to the Movement of a Control Vertex

If an already-existing curve is to be modified, it is not necessary to recompute the entire curve. Careful consideration of the properties of the Beta-spline representation enables an existing curve to be modified in a manner that is more efficient than a complete recomputation.

Consider the consequences to an existing curve when the position of one control vertex is modified. Since a single control vertex influences only four curve segments and has no effect on the other segments, the consequences of moving one vertex are limited to four segments. Computationally, this implies that the movement of a control vertex requires the re-evaluation of only four segments.

Moreover, even the four affected segments need not be completely recomputed. Although each of these segments is controlled by four vertices, only one of these vertices has changed position. Therefore, the change in each of these segments is due only to the modification of the position of one control vertex. Recalling the mathematical formulation for the $i^{th}$ curve segment given in equation (7.2), the change in this segment, $\mathbf{Q}_i(u)$, can be written as

$$\mathbf{Q}_i^A(u) = \sum_{r=-2}^{1} b_r(\beta 1, \beta 2; u)\mathbf{V}_{i+r}^A \tag{8.7}$$

where $\mathbf{V}_i^A$ is the change in position of control vertex $\mathbf{V}_i$.

Denoting the modified control vertex as $\mathbf{V}_{\hat{i}}$, and assuming all the other vertices remain unchanged, then

$$\mathbf{V}_i^A = 0 \qquad\qquad\qquad \text{for } i \neq \hat{i} . \tag{8.8}$$

Thus, only one of the four terms in equation (8.7) is nonzero, and hence this equation reduces to

$$\mathbf{Q}_i^A(u) = b_r(\beta 1, \beta 2; u)\mathbf{V}_{i+r}^A \tag{8.9}$$

where

$$i + r = \hat{i} .$$

Rewriting this equation as

$$\mathbf{Q}_{\hat{i}-r}^A(u) = b_r(\beta 1, \beta 2; u)\mathbf{V}_{\hat{i}}^A \qquad\qquad \text{for } r = -2, -1, 0, 1 , \tag{8.10}$$

it is easily seen that the four affected curve segments are $\mathbf{Q}_i(u)$, where

$$i = \hat{i} - r \qquad\qquad\qquad \text{for } r = -2, -1, 0, 1 . \tag{8.11}$$

Therefore, the change in position of control vertex $\mathbf{V}_{\hat{i}}$ perturbs the segments $\mathbf{Q}_i(u)$ by

$$\mathbf{Q}_i^A(u) = b_{\hat{i}-i}(\beta 1, \beta 2; u)\mathbf{V}_{\hat{i}}^A \tag{8.12}$$

where $i$ takes on the values specified by equation (8.11).

Since equation (8.12) represents the change in the curve segment $\mathbf{Q}_i(u)$, the new segment can be determined by incrementing the old segment by this change:

$$\mathbf{Q}_i^{\text{new}}(u) = \mathbf{Q}_i^{\text{old}}(u) + b_{\hat{i}-i}(\beta 1, \beta 2; u)\mathbf{V}_{\hat{i}}^A . \tag{8.13}$$

To compute the new curve resulting from modifying the position of the control vertex $\mathbf{V}_{\hat{i}}$, equation (8.13) is evaluated for all the necessary values of the parameter $u$ and for each of the four curve segments $\mathbf{Q}_i(u)$ with the values of $i$ given by equation (8.11). Thus, the algorithm to compute the four perturbed curve segments is:

> **for** $i := \hat{i} - 1$ **to** $\hat{i} + 2$ **do**
>> **for each** $u$ **in** $\{u_{i_k} | k = 0, 1, \ldots, p_i\}$ **do**
>> **begin**
>>> compute_current_b $(\hat{i} - i, c, u, b)$;
>>> $\mathbf{Q}_i^{new}(u) := \mathbf{Q}_i^{old}(u) + b_{\hat{i}-i} * \mathbf{V}_{\hat{i}}^{\Delta}$
>> **end**;

where

> **procedure** compute_current_b $(r; c; u; b)$;
> **begin** (∗ compute_current_b ∗)
>> **case** $r$ **of**
>> $-2$:   $b_{-2}(beta1, beta2; u) := c_{0,-2} * (1 - u)^3$;
>> $-1$:   $b_{-1}(beta1, beta2; u) := c_{0,-1} + u * (c_{1,-1} + u * (c_{2,-1} + u * c_{3,-1}))$;
>>  0:   $b_0(beta1, beta2; u) := c_{0,0} + u * (c_{1,0} + u * (c_{2,0} + u * c_{3,0}))$;
>>  1:   $b_1(beta1, beta2; u) := c_{3,1} * u^3$
>> **end** (∗ case ∗)
> **end** (∗ compute_current_b ∗);

This algorithm requires a total of $\sum\limits_{i=\hat{i}-1}^{\hat{i}+2} (p_i + 1)(3 + d)$ multiplications and $(p_{\hat{i}-1} + 1)d + (p_{\hat{i}} + 1)(3 + d) + (p_{\hat{i}+1} + 1)(3 + d) + (p_{\hat{i}+2} + 1)(1 + d)$   additions/subtractions to perturb a general curve where the $i^{th}$ segment is computed at $p_i + 1$ values of the domain parameter.

If the four affected curve segments were computed at the same set of $p + 1$ values of the domain parameters, then these computational requirements would become $(p + 1)(12 + 4d) = 4(p + 1)(3 + d)$ multiplications and $(p + 1)(7 + 4d)$ additions/subtractions. However, the computational requirements can be reduced to $(p + 1)(9 + 4d)$ multiplications and $(p + 1)(9 + 4d)$ additions/subtractions, and the introduction of the compute_current_b algorithm can be avoided as shown in the following algorithm:

> **for each** $u$ **in** $\{u_k | k = 0, 1, \ldots, p\}$ **do**
> **begin**
>> compute_b $(c, u, b)$;
>> **for** $i := \hat{i} - 1$ **to** $\hat{i} + 2$ **do**
>>> $\mathbf{Q}_i^{new}(u) := \mathbf{Q}_i^{old}(u) + b_{\hat{i}-i} * \mathbf{V}_{\hat{i}}^{\Delta}$
> **end**;

In addition, if the basis functions were precomputed for this set of $p + 1$ values of the domain parameter and stored in a table having $4(p + 1)$ storage locations, then a saving of $9(p + 1)$ multiplications and $9(p + 1)$ additions/subtractions could be realized. The following algorithm can be used to compute the four perturbed segments at these parametric values:

**for each** $u$ **in** $\{u_k | k = 0, 1, \ldots, p\}$ **do**

    **for** $i := \hat{i} - 1$ **to** $\hat{i} + 2$ **do**

        $\mathbf{Q}_i^{\text{new}}(u) := \mathbf{Q}_i^{\text{old}}(u) + b_{\hat{i}-i} * \mathbf{V}_{\hat{i}}^{\Delta}$

This algorithm requires one multiplication and one addition for each coordinate of each of the $p + 1$ points on each of the four segments; thus, the total computational requirement is $4d(p + 1)$ multiplications and $4d(p + 1)$ additions.

It should be emphasized that each of these algorithms is based on the assumption that only *one* control vertex has been moved. If the positions of more than one vertex are to be modified, they must be moved one at a time, and the algorithm must be performed for each such change. The algorithms are not valid in cases where the positions of several vertices are modified simultaneously. Also, like all incremental techniques, these calculations could be adversely affected by round-off error.

# 9 Generalizing to Continuous Shape Parameters for Curves

The evaluation and perturbation of a Beta-spline curve has been investigated under the assumption that $\beta 1$ and $\beta 2$ are *uniform* shape parameters. Now they will be generalized to be *continuous* shape parameters, each varying continuously along the curve. The continuous analogues of $\beta 1$ and $\beta 2$ will be denoted $\beta 1_i(u)$ and $\beta 2_i(u)$, respectively, and describe the value of each shape parameter along the curve segment $Q_i(u)$, $i = 1, 2, \ldots, m$. This generalization will enable the user to have more precise control over the shape of the curve. The user is no longer constrained to choose a unique value for each shape parameter over the entire curve. Now, different values of the shape parameters can be used to reflect the local character of the curve.

In order for the mathematical derivation given in Section 7.4 to remain valid, it is necessary that each shape parameter have a unique value at each joint between curve segments; that is, at $Q_{i+1}(0) \, (= Q_i(1))$, $i = 1, 2, \ldots, m - 1$. Specifically,

$$\beta 1_{i+1}(0) = \beta 1_i(1)$$

and                                                                                          (9.1)

$$\beta 2_{i+1}(0) = \beta 2_i(1) \qquad\qquad\qquad\qquad \text{for } i = 1, 2, \ldots, m - 1 \ .$$

The actual value at a joint can be specified by the user as a *discrete* shape parameter. Denoting these discrete analogues by $\alpha 1_i$ and $\alpha 2_i$, the following conditions must be satisfied by $\beta 1_i(u)$ and $\beta 2_i(u)$ (Figure 9.1):

$$\beta 1_1(0) = \alpha 1_0 \qquad\qquad \text{and} \qquad \beta 2_1(0) = \alpha 2_0$$

$$\beta 1_{i+1}(0) = \alpha 1_i = \beta 1_i(1) \quad \text{and} \quad \beta 2_{i+1}(0) = \alpha 2_i = \beta 2_i(1)$$

(9.2)

$$\text{for } i = 1, 2, \ldots, m - 1$$

$$\alpha 1_m = \beta 1_m(1) \qquad\qquad \text{and} \qquad \alpha 2_m = \beta 2_m(1) \ .$$

The following $\beta 1_i(u)$ and $\beta 2_i(u)$ functions are one solution to these conditions. Each function interpolates the value of the discrete shape parameter at the joint, but is parametrized in such a manner that the first and second (scalar) derivatives are zero at the joint. Specifically,

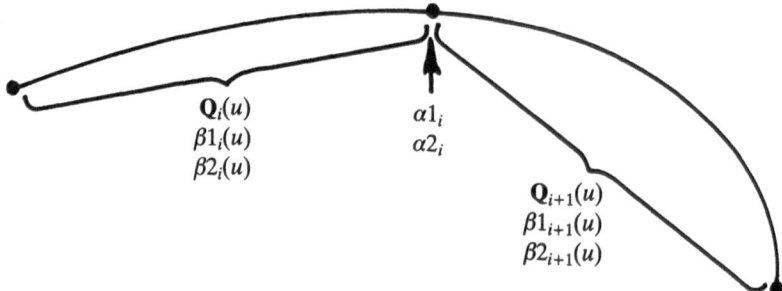

**Fig. 9.1.**   Continuous and discrete shape parameters for curves.

$$\beta 1_i(u) = (1 - s)\, \alpha 1_{i-1} + s\, \alpha 1_i$$

and                                                                                                      (9.3)

$$\beta 2_i(u) = (1 - s)\, \alpha 2_{i-1} + s\, \alpha 2_i$$

where

$$s = 10u^3 - 15u^4 + 6u^5 \qquad\qquad\qquad \text{for } i = 1, 2, \ldots, m \; .$$

In addition, it is necessary to extend the notation

$$\delta = 2\,\beta 1^3 + 4\,\beta 1^2 + 4\,\beta 1 + \beta 2 + 2$$

to

$$\delta_i(u) = 2\,\beta 1_i^3(u) + 4\,\beta 1_i^2(u) + 4\,\beta 1_i(u) + \beta 2_i(u) + 2 \quad \text{for } i = 1, 2, \ldots, m \; . \quad (9.4)$$

Finally, the discrete analogue to $\delta_i(u)$ is

$$\gamma_0 = \delta_1(0)$$

and                                                                                                      (9.5)

$$\gamma_i = \delta_i(1) \qquad\qquad\qquad\qquad\qquad \text{for } i = 1, 2, \ldots, m \; .$$

Now, the original curve representation can be treated as a special case where $\alpha 1_i$ and $\beta 1_i(u)$ are both replaced with $\beta 1$; $\alpha 2_i$ and $\beta 2_i(u)$ are both replaced with $\beta 2$; and $\gamma_i$ and $\delta_i(u)$ are both replaced with $\delta$.

# 10 Curve Evaluation and Perturbation with Continuous Shape Parameters

## 10.1 Evaluation Method I

Since all the coefficient functions have a common denominator of $\delta(u)$, all four basis functions will have this common denominator, and therefore the expression for the curve will have a denominator of $\delta(u)$. It is thus of computational interest to define corresponding sets of coefficient functions and basis functions that are scaled by a factor of $\delta(u)$. This would simplify the expressions and eliminate redundant divisions. These scaled coefficient functions and scaled basis functions will be denoted $d_{gr}(\beta1(u), \beta2(u))$ and $a_r(\beta1(u), \beta2(u); u)$, respectively; in particular,

$$d_{gr}(\beta1(u), \beta2(u)) = \delta(u)c_{gr}(\beta1(u), \beta2(u))$$

$$\text{for } g = 0, 1, 2, 3 \quad \text{and} \quad r = -2, -1, 0, 1 \ , \tag{10.1}$$

and

$$a_r(\beta1(u), \beta2(u); u) = \delta(u)b_r(\beta1(u), \beta2(u); u) \qquad \text{for } r = -2, -1, 0, 1 \ .$$

Multiplying equation (7.3) by $\delta(u)$, and substituting from equation (10.1) yields

$$a_r(\beta1(u), \beta2(u); u) = \sum_{g=0}^{3} d_{gr}(\beta1(u), \beta2(u))u^g \qquad \text{for } r = -2, -1, 0, 1 \ . \tag{10.2}$$

This equation forms the nucleus of the following algorithm which computes the four *scaled* basis functions, given the values of the $d_{gr}$'s for a given value of each shape parameter, $\beta1$ and $\beta2$, and of the domain parameter $u$; it requires 9 multiplications and 9 additions/subtractions.

```
procedure compute_a (d; delta; u; a);
begin (* compute_a *)
    a₁ := 2 * u³;
    a₀ := d₀,₀ + u * (d₁,₀ + u * (d₂,₀ + u * d₃,₀));
    a₋₁ := d₀,₋₁ + u * (d₁,₋₁ + u * (d₂,₋₁ + u * d₃,₋₁));
    a₋₂ := delta − a₁ − a₀ − a₋₁
end (* compute_a *);
```

After the basis functions have been computed, a curve point at the domain parametric value $u$ can be computed by the following algorithm which requires $4d$ multiplications, $3d$ additions/subtractions, and $d$ divisions.

**procedure** compute_Q $(i; a; delta; V; Q)$;

**begin** ($*$ compute_Q $*$)

   $Q_i := a_{-2} * V_{i-2}$;

   **for** $r := -1$ **to** 1 **do** $Q_i := Q_i + a_r * V_{i+r}$;

   $Q_i := Q_i/delta$

**end** ($*$ compute_Q $*$);

Now, evaluation method I described in Section 8.1 can be generalized to *continuous* shape parameters. The following algorithm computes a general curve composed of $m$ segments where the $i^{\text{th}}$ segment is evaluated at $p_i + 1$ values of the domain parameter:

**for** $i := 1$ **to** $m$ **do**

**begin**

   $jump1 := alpha1_i - alpha1_{i-1}$;

   $jump2 := alpha2_i - alpha2_{i-1}$;

   **for each** $u$ **in** $\{u_{i_k} | k = 0, 1, \ldots, p_i\}$ **do**

   **begin**

      $s := (10 + (6 * u - 15) * u) * u * u * u$;

      $beta1 := alpha1_{i-1} + s * jump1$;

      $beta2 := alpha2_{i-1} + s * jump2$;

      compute_d $(beta1, beta2, d)$;

      compute_delta $(d, delta)$;

      compute_a $(d, delta, u, a)$;

      compute_Q $(i, a, delta, V, Q)$

   **end**

**end**;

This algorithm requires $\sum_{i=1}^{m} (p_i + 1)(23 + 4d)$ multiplications, $\sum_{i=1}^{m} (2 + (p_i + 1)$ $(27 + 3d))$ additions/subtractions, and $\sum_{i=1}^{m} d(p_i + 1)$ divisions.

Using the notation defined in equation (8.1), and rearranging, these computational requirements become $m(23 + 4d)(p + 1)$ multiplications, $m(2 + 3(9 + d) \times (p + 1))$ additions/subtractions, and $dm(p + 1)$ divisions. It should be noted that this algorithm naturally computes the general curve evaluated at a different set of values of the domain parameter on each segment, with no loss of efficiency.

## 10.2 Evaluation Method II

Generalizing the equations derived in Section 8.2 to continuous shape parameters and writing them in terms of scaled coefficient functions and scaled basis functions results in

$$Q_i(u) = \frac{1}{\delta(u)} [S_{i0}(\beta 1(u), \beta 2(u)) + u(S_{i1}(\beta 1(u), \beta 2(u))$$

$$+ u(S_{i2}(\beta 1(u), \beta 2(u)) + u(S_{i3}(\beta 1(u), \beta 2(u))))] \tag{10.3}$$

where $S_{ig}(\beta 1(u), \beta 2(u))$ is given by

$$S_{ig}(\beta 1(u), \beta 2(u)) = \sum_{r=-2}^{1} d_{gr}(\beta 1(u), \beta 2(u)) V_{i+r} \qquad \text{for } g = 0, 1, 2, 3 . \tag{10.4}$$

Using these equations, the following algorithm computes a general curve composed of $m$ segments where the $i^{th}$ segment is evaluated at $p_i + 1$ values of the domain parameter:

**for** $i := 1$ **to** $m$ **do**
**begin**
    $jump1 := alpha1_i - alpha1_{i-1};$
    $jump2 := alpha2_i - alpha2_{i-1};$
    **for each** $u$ **in** $\{u_{i_k} | k = 0, 1, \ldots, p_i\}$ **do**
    **begin**
        $s := (10 + (6 * u - 15) * u) * u * u * u;$
        $beta1 := alpha1_{i-1} + s * jump1;$
        $beta2 := alpha2_{i-1} + s * jump2;$
        compute_all_d $(beta1, beta2, d);$
        compute_delta $(d, delta);$
        **for** $g := 0$ **to** 3 **do**
        **begin**
            $S_{ig} := d_{g,-2} * V_{i-2};$
            **for** $r := -1$ **to** 1 **do** $S_{ig} := S_{ig} + d_{gr} * V_{i+r}$
        **end**;
        $Q_i := S_{i3};$
        **for** $g := 2$ **downto** 0 **do** $Q_i := S_{ig} + u * Q_i;$
        $Q_i := Q_i/delta$
    **end**
**end**;

where procedure compute_all_d $(beta1; beta2; d)$ is defined as follows:

**procedure** compute_all_d (*beta*1; *beta*2; *d*);

**begin** (∗ compute_all_d ∗)

$t_i := beta1^2 + beta2;$

$t_2 := beta1^2 + t_1;$

$d_{3,-2} := beta1 * (beta2 - t_2);$

$d_{2,-2} := 3 * d_{3,-2};$

$d_{1,-2} := -d_{2,-2};$

$d_{0,-2} := -d_{3,-2};$

$d_{0,0} := 2;$

$d_{1,0} := 6 * beta1;$

$d_{2,0} := 3 * t_2;$

$d_{3,0} := -2 * (t_1 + beta1 + 1);$

$d_{0,-1} := d_{1,0} + d_{2,0} + d_{3,0} + 2;$

$t_3 := 3 * d_{0,-2};$

$d_{1,-1} := t_3 - d_{1,0};$

$d_{2,-1} := t_3 - d_{2,0};$

$d_{3,-1} := d_{0,-2} - d_{3,0} - 2;$

$d_{3,1} := 2;$

$d_{2,1} := 0;$

$d_{1,1} := 0;$

$d_{0,1} := 0;$

**end** (∗ compute_all_d ∗);

This algorithm requires $\sum_{i=1}^{m} (p_i + 1)(5 + 2 + 8 + d\{4(1 + 3) + 3\})$ multiplications, $\sum_{i=1}^{m} [2 + (p_i + 1)(2 + 2 + 12 + 2 + d\{4(3) + 3\})]$ additions/subtractions, and $\sum_{i=1}^{m} d(p_i + 1)$ divisions. Using the notation defined in equation (8.1), and rearranging, these computational requirements become $m(15 + 19d)(p + 1)$ multiplications, $m(2 + 3(6 + 5d)(p + 1))$ additions/subtractions, and $dm(p + 1)$ divisions.

These computational requirements are slightly more than those for the algorithm designed in Section 10.1. Again, this algorithm computes the general curve evaluated at a different set of values of the domain parameter on each segment, with no loss of efficiency.

## 10.3  Perturbation Due to the Movement of a Control Vertex

Analogous to the perturbation of a Beta-spline curve with uniform shape parameters explained in Section 8.4, an already-existing curve with continuous shape parameters can be modified in a manner that is more efficient than a complete recomputation.

The following algorithm perturbs a general curve where the $i^{th}$ curve segment is computed at $p_i + 1$ values of the domain parameter:

**for** $i := \hat{i} - 1$ **to** $\hat{i} + 2$ **do**
**begin**
  $jump1 := alpha1_i - alpha1_{i-1};$
  $jump2 := alpha2_i - alpha2_{i-1};$
  **for each** $u$ **in** $\{u_{i_k} | k = 0, 1, \ldots, p_i\}$ **do**
  **begin**
    $s := (10 + (6 * u - 15) * u) * u * u * u;$
    $beta1 := alpha1_{i-1} + s * jump1;$
    $beta2 := alpha2_{i-1} + s * jump2;$
    compute_d $(beta1, beta2, d);$
    compute_delta $(d, delta);$
    compute_current_a $(\hat{i} - i, d, u, a);$
    $b_{\hat{i}-i} := a_{\hat{i}-i}/delta;$
    $Q_i^{new}(u) := Q_i^{old}(u) + b_{\hat{i}-i} * V_{\hat{i}}^{\Delta}$
  **end**
**end;**

where

  **procedure** compute_current_a $(r; d; u; a);$
  **begin** ($*$ compute_current_a $*$)
    **case** $r$ **of**
    $-2$:  $a_{-2}(beta1, beta2; u) := d_{0,-2} * (1 - u)^3;$
    $-1$:  $a_{-1}(beta1, beta2; u) := d_{0,-1} + u * (d_{1,-1} + u * (d_{2,-1} + u * d_{3,-1}));$
    $0$:  $a_0(beta1, beta2; u) := d_{0,0} + u * (d_{1,0} + u * (d_{2,0} + u * d_{3,0}));$
    $1$:  $a_1(beta1, beta2; u) := 2 * u^3$
    **end** ($*$ case $*$)
  **end** ($*$ compute_current_a $*$);

For comparison purposes, assume that the four curve segments affected were computed at the same set of $p + 1$ values of the domain parameters. Then the computational requirements of this algorithm could be expressed as $4(p + 1)$ $(5 + 2 + 7 + 3 + d) = 4(p + 1)(17 + d)$ multiplications, $4\left[2 + (p + 1)\left(2 + 2 + 12 + 2 + \frac{7}{4} + d\right)\right] = 8 + (p + 1)(79 + 4d)$ additions/subtractions, and $4(p + 1)$ divisions.

## 10.4  Perturbation Due to the Modification of Shape Parameters

A Beta-spline curve with continuous shape parameters can also be locally modified by altering the values of the shape parameters at a joint. Note that the continuous shape parameters for the $i^{\text{th}}$ curve segment, $\beta 1_i(u)$ and $\beta 2_i(u)$, are functions of $\alpha 1_{i-1}$ and $\alpha 1_i$, and of $\alpha 2_{i-1}$ and $\alpha 2_i$, respectively. Denoting the modified shape parameters as $\alpha 1_{\hat{\imath}}$ and $\alpha 2_{\hat{\imath}}$, the affected curve segments are $\mathbf{Q}_{\hat{\imath}}(u)$ and $\mathbf{Q}_{\hat{\imath}+1}(u)$. Thus, changing the values of the shape parameters at one joint requires the re-evaluation of only two curve segments.

# 11 Classification and Analysis of Beta-spline Curve End Conditions

## 11.1 Introduction

The complete definition of a curve or surface by an open Beta-spline formulation requires the specification of an *end* condition in the case of a curve, or a *boundary* condition for a surface. Different techniques have various geometric properties which require careful study to enable the selection of an appropriate approach. A classification and explanation of various end conditions and boundary conditions for uniform B-spline curve and surface representations, as well as an analysis of their geometric properties, was previously presented by the author in [3]. Analogous discussions for the Beta-spline curve and surface representations are presented in this section and Chapter 16, respectively.

## 11.2 Classification

Recall that the control polygon is composed of the $m + 1$ control vertices

$$\{V_0, V_1, \ldots, V_m\}$$

(Figure 2.1). Considering the Beta-spline curve formulation (equation (7.2)), it can be seen that these vertices can be used to generate $m - 2$ curve segments, specifically $Q_2(u), Q_3(u), \ldots, Q_{m-1}(u)$ (Figure 11.1).

Note that the Beta-spline curve starts at

$$Q_2(0) = \frac{2\,\alpha 1_1^3}{\gamma_1} V_0 + \left[1 - 2\frac{\alpha 1_1^3 + 1}{\gamma_1}\right] V_1 + \frac{2}{\gamma_1} V_2$$

and ends at

$$Q_{m-1}(1) = \frac{2\,\alpha 1_{m-1}^3}{\gamma_{m-1}} V_{m-2} + \left[1 - 2\frac{\alpha 1_{m-1}^3 + 1}{\gamma_{m-1}}\right] V_{m-1} + \frac{2}{\gamma_{m-1}} V_m \ .$$

To have the curve start closer to $V_0$ and end nearer $V_m$, additional curve segments can be defined at the ends. The definition of these segments, however, cannot be done by evaluating equation (7.2) in the usual way since this would reference non-existent vertices. Various methods are available for defining these curve segments

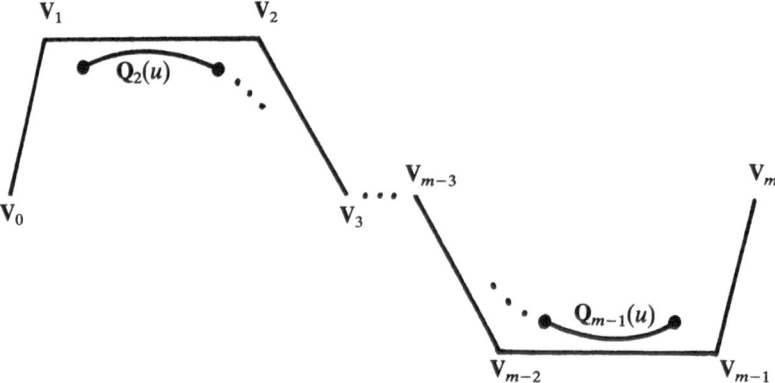

**Fig. 11.1.**   Interior segments naturally defined by control polygon.

and will now be described. These techniques fall into two classifications, *multiple vertices* and *phantom vertices*.

## 11.3   Analysis of Multiple Vertices End Conditions

### 11.3.1   Double Vertices

The double vertices technique defines one additional curve segment at each end by repeating the end vertex in the Beta-spline curve formulation. This technique yields a Beta-spline curve composed of $m$ segments, that is $Q_1(u)$, $Q_2(u)$, ..., $Q_m(u)$. Since the control polygon has $m$ line segments, this is a corresponding number of curve segments.

The additional segments are $Q_1(u)$ and $Q_m(u)$ (Figure 11.2) and they are defined by equation (7.2) in the usual manner except that vertices $V_0$ and $V_m$ are used when $V_{-1}$ and $V_{m+1}$, respectively, are referenced. Thus,

$$Q_1(u) = [b_{-2}(u) + b_{-1}(u)]V_0 + b_0(u)V_1 + b_1(u)V_2 \tag{11.1}$$

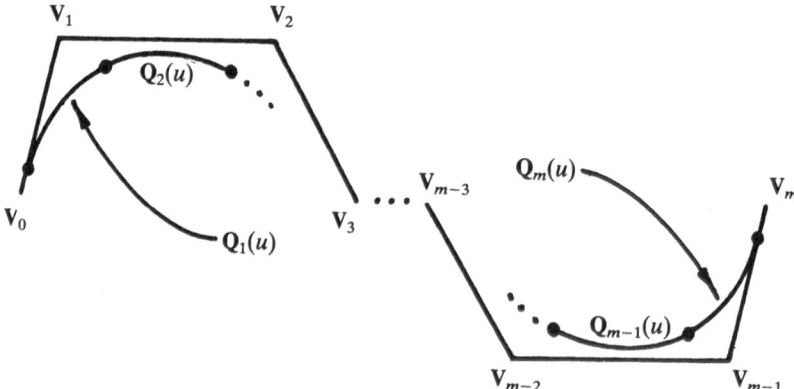

**Fig. 11.2.**   Additional segments from double vertices end condition.

and

$$Q_m(u) = b_{-2}(u)V_{m-2} + b_{-1}(u)V_{m-1} + [b_0(u) + b_1(u)]V_m \ . \tag{11.2}$$

With these additional curve segments, the Beta-spline curve now starts at

$$Q_1(0) = \left(1 - \frac{2}{\gamma_0}\right)V_0 + \frac{2}{\gamma_0}V_1$$

and ends at

$$Q_m(1) = \frac{2\,\alpha 1_m^3}{\gamma_m}V_{m-1} + \left(1 - \frac{2\,\alpha 1_m^3}{\gamma_m}\right)V_m \ .$$

Thus, the initial point of the curve is $\dfrac{2}{\gamma_0}$ along the vector from $V_0$ to $V_1$ and the terminal point is $\left[1 - \dfrac{2\,\alpha 1_m^3}{\gamma_m}\right]$ along the vector from $V_{m-1}$ to $V_m$.

At these endpoints, the curve is tangent to the control polygon. To show this, the first derivative vector can be evaluated by taking the first derivative of equation (11.1), evaluating at $u = 0$ and $u = 1$, and substituting the values of the first derivative of each basis function (as provided in Section 7.4), yielding:

$$Q_1^{(1)}(0) = \frac{6\,\alpha 1_0}{\gamma_0}[V_1 - V_0]$$

$$Q_m^{(1)}(1) = \frac{6\,\alpha 1_m^2}{\gamma_m}[V_m - V_{m-1}] \ . \tag{11.3}$$

An important property of an end condition is the curvature at the endpoints of the curve. It is sometimes desirable to have zero curvature at an endpoint to join the curve to a straight line, while other applications require nonzero curvature.

Sufficient conditions for zero or nonzero curvature were given in Section 4.5.4, and required the evaluation of the first and second derivative vectors. The value of the second derivative vector at each endpoint of the curve is found in a manner analogous to that which was used for the first derivative where the values of the second derivative of each basis function are obtained by consulting Section 7.4:

$$Q_1^{(2)}(0) = \frac{6(2\,\alpha 1_0^2 + \alpha 2_0)}{\gamma_0}[V_1 - V_0]$$

$$Q_m^{(2)}(1) = \frac{6(2\,\alpha 1_m + \alpha 2_m)}{\gamma_m}[V_{m-1} - V_m] \ . \tag{11.4}$$

Comparing the above expressions for the first and second derivative vectors at each endpoint of the curve,

$$Q_1^{(2)}(0) = \frac{2\,\alpha 1_0^2 + \alpha 2_0}{\alpha 1_0}Q_1^{(1)}(0)$$

$$Q_m^{(2)}(1) = -\frac{2\,\alpha 1_m + \alpha 2_m}{\alpha 1_m^2}Q_m^{(1)}(1) \ . \tag{11.5}$$

Assuming $V_0$ and $V_1$ are distinct, then the first derivative is nonzero at the initial point of the curve and the first and second derivative vectors there are linearly dependent; this is sufficient to conclude, from Condition 1 of Section 4.5.4, that the curvature there is zero. A similar process will show that the curvature is also zero at the terminal point of the curve.

Finally, it should also be noted that the curve segments defined by this technique are, in general, smaller than those defined by the unmodified Beta-spline curve formulation. This is due to the fact that the double vertex assumes the role of two vertices; thus, the vertices that define such a curve segment are, in general, more clustered than those which define a regular curve segment.

### 11.3.2 Triple Vertices

The triple vertices technique is an extension of the double vertices technique. In fact, it is simply the double vertices technique with the definition of another additional curve segment at each end. The segments $Q_1(u)$ and $Q_m(u)$ are defined by the double vertices technique and the segments $Q_0(u)$ and $Q_{m+1}(u)$ (Figure 11.3) are defined by equation (7.2) using $V_0$ whenever $V_{-1}$ or $V_{-2}$ is referenced and using $V_m$ for $V_{m+1}$ or $V_{m+2}$. Thus,

$$Q_0(u) = [b_{-2}(u) + b_{-1}(u) + b_0(u)]V_0 + b_1(u)V_1$$

$$Q_{m+1}(u) = b_{-2}(u)V_{m-1} + [b_{-1}(u) + b_0(u) + b_1(u)]V_m .$$

$$(11.6)$$

Substituting the expressions for the basis functions yields

$$Q_0(u) = \left[1 - \frac{2u^3}{\delta_0(u)}\right]V_0 + \frac{2u^3}{\delta_0(u)}V_1$$

$$Q_{m+1}(u) = \frac{2(\beta 1_{m+1}(u) * (1-u))^3}{\delta_{m+1}(u)}V_{m-1}$$

$$(11.7)$$

$$+ \left[1 - \frac{2(\beta 1_{m+1}(u) * (1-u))^3}{\delta_{m+1}(u)}\right]V_m .$$

As $u$ varies from 0 to 1, $Q_0(u)$ traces a straight line segment starting at $V_0$ and

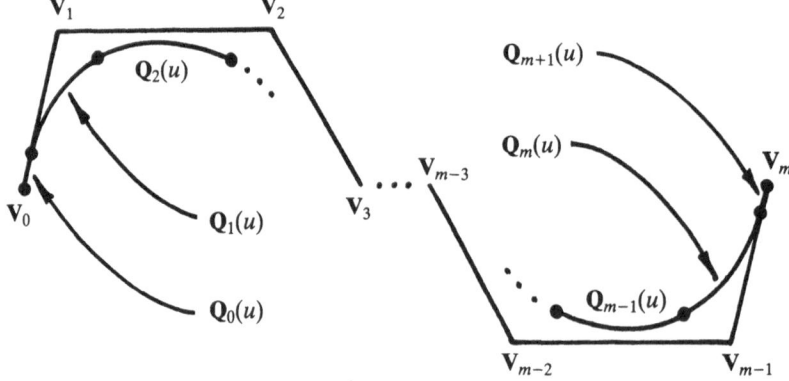

**Fig. 11.3.**  Additional segments from triple vertices end condition.

terminating at the point which is $\dfrac{2}{\gamma_0}$ along the vector to $V_1$, while $Q_{m+1}(u)$ traces

a line segment whose initial point is $\left[1 - \dfrac{2\,\alpha 1_m^3}{\gamma_m}\right]$ along the vector from $V_{m-1}$

to $V_m$ and whose terminal point is $V_m$. Note that this technique has the advantage that the control vertex at each end is interpolated.

Although these curve segments are straight line segments, it is important to

observe that they are, in general, quite short since $Q_0(u)$ has a length of $\dfrac{2}{\gamma_0}$ times

the distance between $V_0$ and $V_1$ and $Q_m(u)$ has a length of $\dfrac{2\,\alpha 1_m^3}{\gamma_m}$ times the distance

between $V_{m-1}$ and $V_m$. Although the vertices are free to be located anywhere and thus this length is unconstrained, these factors mean that the length will, in general, be quite small.

## 11.4 Analysis of Phantom Vertices End Conditions

### 11.4.1 Description

With these techniques, an auxiliary vertex is created at each end of the control polygon. This can then be used to define an additional curve segment at each end by evaluating the Beta-spline curve formulation (equation (7.2)) in the same manner as for the curve segments defined by the original control polygon (Figure 11.4).

The auxiliary vertices are created for the sole purpose of defining the additional curve segments, and are inaccessible to the user and not displayed; thus, they will be referred to as *phantom vertices*. The phantom vertices are completely defined in terms of the original control vertices in such a manner as to satisfy some end condition. Several such end conditions are discussed in the following sections.

Since both additional curve segments are defined by an *unmodified* Beta-spline curve formulation, they are of normal size. Further, the phantom vertices techniques generate curves consisting of $m$ segments, which is a convenient number of segments since the control polygon has $m$ line segments.

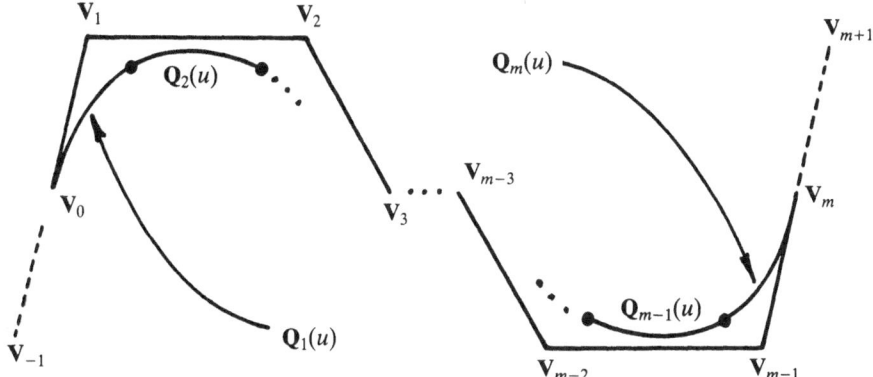

**Fig. 11.4.** The phantom vertices and additional curve segments defined by phantom vertices end condition.

This *phantom vertices* concept was developed independently of that mentioned by Coons in [7], although the underlying idea is similar. The latter case was developed only for B-spline curves and not for surfaces, and this description of a curve involved a total of *four* phantom vertices: two *initiating* vertices at the beginning of the curve, and two *terminating* vertices at the end. The two initiating vertices are defined in terms of the first non-phantom vertex and two furnished points. Similarly, the two phantom terminating vertices are expressed in terms of the final non-phantom vertex and another two specified points.

### 11.4.2 Position Specification

Using this end condition, the phantom vertices are defined such that each endpoint of the curve interpolates a specified point. That is,

$$
\begin{aligned}
\mathbf{Q}_1(0) &= \mathbf{P}_0 \\
\mathbf{Q}_m(1) &= \mathbf{P}_m \; .
\end{aligned}
\tag{11.8}
$$

Evaluating the left-hand sides of these equations by substituting the extreme parametric values 0 and 1 into equation (7.2) and solving for the phantom vertices results in

$$
\begin{aligned}
\mathbf{V}_{-1} &= \frac{1}{2\,\alpha 1_0^3}(\gamma_0 \mathbf{P}_0 - [(\gamma_0 - 2\,\alpha 1_0^3 - 2)\mathbf{V}_0 + 2\mathbf{V}_1]) \\
\mathbf{V}_{m+1} &= \frac{1}{2}(\gamma_m \mathbf{P}_m - [(\gamma_m - 2\,\alpha 1_m^3 - 2)\mathbf{V}_m + 2\,\alpha 1_m^3 \mathbf{V}_{m-1}]) \; .
\end{aligned}
\tag{11.9}
$$

To analyze the curvature at the endpoints of the curve, the first derivative vector at each endpoint of the curve is evaluated by taking the first derivative of equation (7.2) and substituting the appropriate values of the first derivative of the basis functions:

$$
\begin{aligned}
\mathbf{Q}_1^{(1)}(0) &= \frac{6\,\alpha 1_0}{\gamma_0}[-\alpha 1_0^2 \mathbf{V}_{-1} + (\alpha 1_0^2 - 1)\mathbf{V}_0 + \mathbf{V}_1] \\
\mathbf{Q}_m^{(1)}(1) &= \frac{6}{\gamma_m}[-\alpha 1_m^2 \mathbf{V}_{m-1} + (\alpha 1_m^2 - 1)\mathbf{V}_m + \mathbf{V}_{m+1}] \; .
\end{aligned}
\tag{11.10}
$$

Substituting the expressions for the phantom vertices given in equation (11.9) into equation (11.10) results in the following expressions for the first derivative vector at each endpoint of the curve:

$$
\begin{aligned}
\mathbf{Q}_1^{(1)}(0) &= 3\left[\frac{2(\alpha 1_0 + 1)}{\gamma_0}[\mathbf{V}_1 - \mathbf{V}_0] + \mathbf{V}_0 - \mathbf{P}_0\right] \\
\mathbf{Q}_m^{(1)}(1) &= 3\left[\frac{2\,\alpha 1_m^2(\alpha 1_m + 1)}{\gamma_m}[\mathbf{V}_m - \mathbf{V}_{m-1}] - \mathbf{V}_m + \mathbf{P}_m\right] \; .
\end{aligned}
\tag{11.11}
$$

Analogously, the value of the second derivative vector at each endpoint of the curve is

$$Q_1^{(2)}(0) = \frac{6}{\gamma_0} [2\, \alpha 1_0^3 \mathbf{V}_{-1} - (2\, \alpha 1_0^3 + 2\, \alpha 1_0^2 + \alpha 2_0)\mathbf{V}_0 + (2\, \alpha 1_0^2 + \alpha 2_0)\mathbf{V}_1]$$

$$(11.12)$$

$$Q_m^{(2)}(1) = \frac{6}{\gamma_m} [(2\, \alpha 1_m + \alpha 2_m)\mathbf{V}_{m-1} - (2\, \alpha 1_m + \alpha 2_m + 2)\mathbf{V}_m + 2\mathbf{V}_{m+1}] \ .$$

Performing a similar substitution shows that the second derivative vector at each endpoint of the curve can be written as

$$Q_1^{(2)}(0) = 6\left[\frac{2\, \alpha 1_0^2 + \alpha 2_0 - 2}{\gamma_0}[\mathbf{V}_1 - \mathbf{V}_0] - \mathbf{V}_0 + \mathbf{P}_0\right]$$

$$(11.13)$$

$$Q_m^{(2)}(1) = 6\left[\frac{-2\, \alpha 1_m^3 + 2\, \alpha 1_m + \alpha 2_m}{\gamma_m}[\mathbf{V}_{m-1} - \mathbf{V}_m] - \mathbf{V}_m + \mathbf{P}_m\right] \ .$$

Comparing equations (11.11) to (11.13), it can be seen that the first and second derivative vectors at each endpoint of the curve are, in general, linearly independent. Thus, the curvature is nonzero at each endpoint of the curve by Condition 2 of Section 4.5.4.

### 11.4.3 End Vertex Interpolation

Although the freedom to select the initial and terminal position of the curve is frequently desirable, it is often convenient to constrain these positions to coincide with the initial and terminal vertex, respectively; that is, to have the curve start at $\mathbf{V}_0$ and end at $\mathbf{V}_m$. This is a special case of the previous end condition where $\mathbf{P}_0 = \mathbf{V}_0$ and $\mathbf{P}_m = \mathbf{V}_m$.

Expressions for the phantom vertices can be obtained by substituting $\mathbf{P}_0 = \mathbf{V}_0$ and $\mathbf{P}_m = \mathbf{V}_m$ in equation (11.9) yielding

$$\mathbf{V}_{-1} = \frac{1}{\alpha 1_0^3}[\mathbf{V}_0 - \mathbf{V}_1] + \mathbf{V}_0$$

$$(11.14)$$

$$\mathbf{V}_{m+1} = \alpha 1_m^3 [\mathbf{V}_m - \mathbf{V}_{m-1}] + \mathbf{V}_m \ .$$

The value of the first derivative vector at each endpoint is given by equation (11.11) with $\mathbf{P}_0 = \mathbf{V}_0$ and $\mathbf{P}_m = \mathbf{V}_m$. Making this substitution into equation (11.11) yields

$$Q_1^{(1)}(0) = \frac{6(\alpha 1_0 + 1)}{\gamma_0}[\mathbf{V}_1 - \mathbf{V}_0]$$

$$(11.15)$$

$$Q_m^{(1)}(1) = \frac{6\, \alpha 1_m^2 (\alpha 1_m + 1)}{\gamma_m}[\mathbf{V}_m - \mathbf{V}_{m-1}] \ .$$

Note that this shows that the curve is tangent to the control polygon at each endpoint.

Making the same substitution into equation (11.13) yields

$$Q_1^{(2)}(0) = \frac{6(2\, \alpha 1_0^2 + \alpha 2_0 - 2)}{\gamma_0}[\mathbf{V}_1 - \mathbf{V}_0]$$

$$(11.16)$$

$$Q_m^{(2)}(1) = \frac{6(2\, \alpha 1_m^3 - 2\, \alpha 1_m - \alpha 2_m)}{\gamma_m}[\mathbf{V}_m - \mathbf{V}_{m-1}] \ .$$

Comparing the above expressions for the first and second derivative vectors at the initial point of the curve,

$$Q_1^{(2)}(0) = \frac{2\,\alpha1_0^2 + \alpha2_0 - 2}{\alpha1_0 + 1}\,Q_1^{(1)}(0)$$

$$Q_m^{(2)}(1) = \frac{2\,\alpha1_m^3 - 2\,\alpha1_m - \alpha2_m}{\alpha1_m^2(\alpha1_m + 1)}\,Q_m^{(1)}(1)\ .$$

(11.17)

Again assuming distinct vertices, the first derivative is nonzero, and the first and second derivative vectors are linearly dependent at the initial and terminal points of the curve. This is sufficient to conclude, from Condition 1 of Section 4.5.4, that the curvature is, in general, zero at each endpoint of the curve.

### 11.4.4  Parametric First Derivative Vector Specification

This end condition defines the phantom vertices by setting the parametric first derivative vector, at each end, equal to some specified value. Specifically,

$$Q_1^{(1)}(0) = P_0^1$$

$$Q_m^{(1)}(1) = P_m^1\ .$$

(11.18)

The left-hand side of these equations can be evaluated using the first derivative of equation (7.2), and then the phantom vertices are determined as

$$V_{-1} = \frac{1}{\alpha1_0^3}[(\alpha1_0^3 - 1)V_0 + V_1 + \frac{\gamma_0}{6}P_0^1]$$

(11.19)

$$V_{m+1} = \alpha1_m^2 V_{m-1} + (1 - \alpha1_m^2)V_m + \frac{\gamma_m}{6}P_m^1\ .$$

Substituting the expressions for the phantom vertices given in equation (11.19) into equation (11.12) shows that the second derivative vector at each endpoint of the curve can be written as

$$Q_1^{(2)}(0) = \frac{6(2\alpha1_0^2 + 2\alpha1_0 + \alpha2_0)}{\gamma_0}[V_1 - V_0] - 2P_0^1$$

(11.20)

$$Q_m^{(2)}(1) = \frac{6(2\alpha1_m^2 + 2\alpha1_m + \alpha2_m)}{\gamma_m}[V_{m-1} - V_m] + 2P_m^1\ .$$

Since the value of the first derivative vector at each endpoint of the curve is free to be specified without restriction, it can be seen that the first and second derivative vectors at each endpoint of the curve are, in general, linearly independent; thus, the curvature there is nonzero by Condition 2 of Section 4.5.4.

The endpoints of the curve can be determined using equation (7.2) and substituting the expressions for the phantom vertices given in equation (11.19):

$$Q_1(0) = 2\frac{\alpha1_0 + 1}{\gamma_0}[V_1 - V_0] + V_0 - \frac{1}{3}P_0^1$$

(11.21)

$$Q_m(1) = 2\alpha1_m^2(\alpha1_m + 1)[V_{m-1} - V_m] + V_m + \frac{1}{3}P_m^1\ .$$

### 11.4.5 Parametric Second Derivative Vector Specification

The phantom vertices are defined by this end condition by setting the parametric second derivative vector, at each end, equal to some specified value. In particular,

$$Q_1^{(2)}(0) = P_0^2$$
$$Q_m^{(2)}(1) = P_m^2 \ .$$
(11.22)

Evaluating the left-hand side of these equations and solving for the phantom vertices yields

$$V_{-1} = \frac{2\,\alpha 1_0^2 + \alpha 2_0}{2\alpha 1_0^3}[V_0 - V_1] + V_0 + \frac{\gamma_0}{12\,\alpha 1_0^3}P_0^2$$

$$V_{m+1} = \frac{2\,\alpha 1_m + \alpha 2_m}{2}[V_m - V_{m-1}] + V_m + \frac{\gamma_m}{12}P_m^2 \ .$$
(11.23)

Substituting the expressions for the phantom vertices given in equation (11.23) into equation (11.10) shows that the first derivative vector at each endpoint of the curve is:

$$Q_1^{(1)}(0) = 3\frac{2\,\alpha 1_0^2 + 2\,\alpha 1_0 + \alpha 2_0}{\gamma_0}[V_1 - V_0] - \frac{1}{2}P_0^2$$

$$Q_m^{(1)}(1) = 3\frac{2\,\alpha 1_m^2 + 2\,\alpha 1_m + \alpha 2_m}{\gamma_m}[V_m - V_{m-1}] + \frac{1}{2}P_m^2 \ .$$
(11.24)

Since the value of the second derivative vector at each endpoint of the curve is free to be specified without restriction, it can be seen that the first and second derivative vectors at each endpoint of the curve are, in general, linearly independent; thus, the curvature there is nonzero by Condition 2 of Section 4.5.4.

The endpoints of the curve can be determined using equation (7.2) and substituting the expressions for the phantom vertices given in equation (11.23):

$$Q_1(0) = \frac{2\,\alpha 1_0^2 + \alpha 2_0 - 2}{\gamma_0}[V_0 - V_1] + V_0 + \frac{1}{6}P_0^2$$

$$Q_m(1) = \frac{2\,\alpha 1_m^3 - 2\,\alpha 1_m - \alpha 2_m}{\gamma_m}[V_{m-1} - V_m] + V_m + \frac{1}{6}P_m^2 \ .$$
(11.25)

### 11.4.6 Zero Parametric Second Derivative Vector

Although the capability of specifying the parametric second derivative vector at each end of the curve is often desirable, it is frequently convenient to have them set to zero by default. All the results of Section 11.45 remain valid with the substitution:

$$P_0^2 = 0$$
$$P_m^2 = 0 \ .$$
(11.26)

However, the zero value of the parametric second derivative vector means that the first and second derivative vectors at each endpoint of the curve are now trivially linearly dependent; thus, the curvature there is zero by Condition 1 of Section 4.5.4.

## 11.5  Conclusion

End conditions for a Beta-spline curve have been classified into two categories, *multiple vertices* and *phantom vertices*. The multiple vertices end conditions that were considered were *double vertices* and *triple vertices*.

The double vertices technique defines one additional curve segment at each end yielding a Beta-spline curve composed of *m* segments. Although the curve does not interpolate the end vertices of the polygon, it is tangent to the control polygon, with zero curvature, at its endpoints. Also, each additional curve segment is, in general, smaller than a regular segment.

The triple vertices technique extends the double vertices technique by defining another additional curve segment at each end. It was shown that these additional segments are actually short straight line segments that are tangent to the control polygon and connect the end vertex to the endpoint of the curve defined by the double vertices technique.

The phantom vertices technique creates an auxiliary vertex at each end of the control polygon to define an additional curve segment at each end. These segments are of normal size and the resulting curve consists of *m* segments. The *phantom vertices* are completely defined in terms of the original vertices such that some end condition, several of which have been discussed, is satisfied. *Position specification* defines the phantom vertices such that the curve interpolates a furnished point with nonzero curvature at each endpoint. *End vertex interpolation* is a special case of position specification where the endpoints of the curve coincide with end vertices and now have zero curvature. *Parametric first derivative vector specification* sets the first derivative vector, at each end, to some specified value. The resulting curve, in general, has nonzero curvature at its endpoints. *Parametric second derivative vector specification* sets the second derivative vector to a specified value, at each end, and also generates a curve with nonzero curvature at the endpoints. The *zero parametric second derivative vector* end condition is a special case of parametric second derivative vector specification where the curvature at each endpoint is zero.

# 12  Explanation of the Surface Representation

A point on the $(i, j)^{\text{th}}$ Beta-spline surface patch is a weighted average of the sixteen vertices $V_{i+r, j+s}$, $r = -2, -1, 0, 1$ and $s = -2, -1, 0, 1$. The mathematical formulation for the patch $Q_{ij}(u, v)$ is then

$$Q_{ij}(u, v) = \sum_{r=-2}^{1} \sum_{s=-2}^{1} bb_{rs}(\beta 1, \beta 2; u, v) V_{i+r, j+s} \qquad (12.1)$$

$$\text{for } 0 \leqslant u < 1 \quad \text{and} \quad 0 \leqslant v < 1 .$$

The set of bivariate Beta-spline basis functions is the tensor product of the set of univariate basis functions. That is,

$$bb_{rs}(\beta 1, \beta 2; u, v) = b_r(\beta 1, \beta 2; u) b_s(\beta 1, \beta 2; v) \qquad (12.2)$$

$$\text{for } r = -2, -1, 0, 1 \quad \text{and} \quad s = -2, -1, 0, 1 .$$

Therefore, this formulation can be rewritten as

$$Q_{ij}(u, v) = \sum_{r=-2}^{1} \sum_{s=-2}^{1} b_r(\beta 1, \beta 2; u) V_{i+r, j+s} b_s(\beta 1, \beta 2; v) \qquad (12.3)$$

$$\text{for } 0 \leqslant u < 1 \quad \text{and} \quad 0 \leqslant v < 1 .$$

Observing that $b_r(\beta 1, \beta 2; u)$ is independent of $s$, it can be treated as a constant multiplier in the inner sum; thus, equation (12.3) can be rewritten in the following form:

$$Q_{ij}(u, v) = \sum_{r=-2}^{1} [b_r(\beta 1, \beta 2; u) \sum_{s=-2}^{1} V_{i+r, j+s} b_s(\beta 1, \beta 2; v)] \qquad (12.4)$$

$$\text{for } 0 \leqslant u < 1 \quad \text{and} \quad 0 \leqslant v < 1 .$$

Recall from Chapter 7 that all four univariate basis functions have a denominator of $\delta$. Replacing the basis functions with their scaled counterparts defined in equation (10.1) yields

$$Q_{ij}(u, v) = \frac{1}{\delta^2} \sum_{r=-2}^{1} [a_r(\beta 1, \beta 2; u) \sum_{s=-2}^{1} V_{i+r, j+s} a_s(\beta 1, \beta 2; v)] \qquad (12.5)$$

$$\text{for } 0 \leqslant u < 1 \quad \text{and} \quad 0 \leqslant v < 1 .$$

These mathematical formulations are used in the design of algorithms to construct the Beta-spline surface as explained in the following chapters.

# 13 Surface Evaluation and Perturbation with Uniform Shape Parameters

## 13.1 Evaluation Method I

A Beta-spline surface patch is described by equation (12.4) as the domain parameters $u$ and $v$ both vary continuously from 0 to 1. To display this patch involves the computation of points on the surface for many different values of the domain parameters. The determination of a point on the patch requires the evaluation of the surface formulation at an appropriate $(u, v)$ value. This entails the evaluation of the four basis functions at the value of $u$ and of $v$, and then the computation of the sum which requires 20 multiplications and 15 additions/subtractions for each coordinate.

The evaluation of the surface points can be accomplished efficiently by exploiting the tensor product form of the Beta-spline surface formulation. Observe that a point on the surface can be thought of as a point on a Beta-spline curve defined by an appropriate set of intermediate control vertices. Specifically, equation (12.4) can be rewritten as

$$Q_{ij}(u, v) = \sum_{r=-2}^{1} b_r(\beta 1, \beta 2; u)W_{i+r,j}(\beta 1, \beta 2; v) \tag{13.1}$$

where

$$W_{i+r,j}(\beta 1, \beta 2; v) = \sum_{s=-2}^{1} V_{i+r,j+s}b_s(\beta 1, \beta 2; v) \ .$$

Consider the following algorithm which constructs a general Beta-spline surface composed of $m$ by $n$ patches where the $(i, j)^{\text{th}}$ patch is evaluated at $p_i + 1$ by $q_j + 1$ values of the domain parameters. It is based on the mathematical formulation given in equation (13.1) and $\beta 1$ and $\beta 2$ have been assumed to be *uniform* shape parameters, each assuming a single value over the entire surface.

> compute_d (*beta1, beta2, d*);
> compute_delta (*d, delta*);
> compute_c (*delta, d, c*);
> **for** $j := 1$ **to** $n$ **do**

> **for each** $v$ **in** $\{v_{j_k}|k = 0, 1, \ldots, q_j\}$ **do**
>> compute_b $(c, v, b)$;
>> **for** $i := 1$ **to** $m$ **do**
>> **begin**
>>> **for** $r := -2$ **to** $1$ **do** compute_W $(i, r, j, b, \mathbf{V}, \mathbf{W})$;
>>> **for each** $u$ **in** $\{u_{i_k}|k = 0, 1, \ldots, p_i\}$ **do**
>>> **begin**
>>>> compute_b $(c, u, b)$;
>>>> compute_Q $(i, j, b, \mathbf{W}, \mathbf{Q})$
>>> **end**
>> **end**;

where

> **procedure** compute_W $(i; r; j; b; \mathbf{V}; \mathbf{W})$;
>
> **begin** $(*$ compute_W $*)$
>
> $\mathbf{W}_{i+r,j} := \mathbf{V}_{i+r,j-2} * b_{-2}$;
>
> **for** $s := -1$ **to** $1$ **do** $\mathbf{W}_{i+r,j} := \mathbf{W}_{i+r,j} + \mathbf{V}_{i+r,j+s} * b_s$
>
> **end** $(*$ compute_W $*)$;

and

> **procedure** compute_Q $(i; j; b; \mathbf{W}; \mathbf{Q})$;
>
> **begin** $(*$ compute_Q $*)$
>
> $\mathbf{Q}_{ij} := b_{-2} * \mathbf{W}_{i-2,j}$;
>
> **for** $r := -1$ **to** $1$ **do** $\mathbf{Q}_{ij} := \mathbf{Q}_{ij} + b_r * \mathbf{W}_{i+r,j}$
>
> **end** $(*$ compute_Q $*)$;

The compute_W algorithm requires $3(1 + 3) = 12$ multiplications and $3(3) = 9$ additions/subtractions and the compute_Q algorithm requires $3(1 + 3) = 12$ multiplications and $3(3) = 9$ additions/subtractions. The complete algorithm to construct the entire surface requires a total of $7 + \sum_{j=1}^{n} (q_j + 1)(9 + \sum_{i=1}^{m} [4(12) + (p_i + 1)(9 + 12)])$ multiplications, $12 + 2 + \sum_{j=1}^{n} (q_j + 1)(9 + \sum_{i=1}^{m} [4(9) + (p_i + 1)(9 + 9)])$ additions/subtractions, and 8 divisions. Using the notation defined in equation (8.1), and rearranging, these computational requirements become $7 + 3n(q + 1)(7mp + 23m + 3)$ multiplications, $14 + 9n(q + 1)(2mp + 6m + 1)$ additions/subtractions, and 8 divisions.

Analogous to Beta-spline curve evaluation methods, a more efficient algorithm would be possible if all surface patches were evaluated at the same set of values of the domain parameters. Again, this does not impose any restriction on the selection of a particular set of values of the domain parameters. The following algorithm requires $7 + 3(q + 1)[3 + mn(7p + 23)]$ multiplications, $14 + 9(q + 1)[1 + 2mn(p + 3)]$ additions/subtractions, and 8 divisions to evaluate the surface at $(p + 1)(q + 1)$ parametric values on each of the $mn$ patches.

```
compute_d (beta1, beta2, d);
compute_delta (d, delta);
compute_c (delta, d, c);
for each v in {v₀, v₁, ..., vq} do
begin
   compute_b (c, v, b);
   for i := 1 to m do
   for j := 1 to n do
   begin
      for r := -2 to 1 do compute_W (i, r, j, b, V, W);
      for each u in {u₀, u₁, ..., up} do
      begin
         compute_b (c, u, b);
         compute_Q (i, j, b, W, Q)
      end
   end
end;
```

These computational requirements can be further reduced at the expense of $4(p + 1 + q + 1)$ storage locations. Note that the four basis functions at the $p + 1$ values of the domain parameter $u$ are redundantly computed $mn(q + 1)$ times. If this evaluation were precomputed and stored in a table, a saving of $9(p + 1) \cdot (mn(q + 1) - 1)$ multiplications and $9(p + 1)(mn(q + 1) - 1)$ additions/subtractions could be realized. The following algorithm can be used to evaluate the surface:

```
compute_d (beta1, beta2, d);
compute_delta (d, delta);
compute_c (delta, d, c);
for each u in {u₀, u₁, ..., up} do compute_b (c, u, bu);
for each v in {v₀, v₁, ..., vq} do compute_b (c, v, bv);
for i := 1 to m do
   for j := 1 to n do
      for each v in {v₀, v₁, ..., vq} do
      begin
         for r := -2 to 1 do compute_W (i, r, j, bv, V, W);
         for each u in {u₀, u₁, ..., up} do compute_Q (i, j, bu, W, Q)
      end;
```

This algorithm would require $7 + 9(p + 1 + q + 1) + mn(q + 1)[4(12) + (p + 1)12] = 7 + 3[3(p + q + 2) + 4mn(p + 5)(q + 1)]$ multiplications, $14 + 9 \cdot (p + 1 + q + 1) + mn(q + 1)[4(9) + (p + 1)9] = 14 + 9[(p + q + 2) + mn(p + 5) \cdot (q + 1)]$ additions/subtractions, and 9 divisions to evaluate the surface at $(p + 1) \cdot (q + 1)$ parametric values on each of the $mn$ patches.

It should also be noted that these computational requirements are not symmetrical in $m$ and $n$ or $p$ and $q$. If an application requires disparate values, care should be taken to write the algorithm so as to optimize the computational requirements relative to these values.

## 13.2  Evaluation Method II

Substituting equation (7.3) into equation (13.1) yields

$$Q_{ij}(u, v) = \sum_{r=2}^{1}\left( \sum_{g=0}^{3} c_{gr}(\beta 1, \beta 2)u^g \right) W_{i+r, j}(\beta 1, \beta 2; v) \qquad (13.2)$$

where

$$W_{i+r, j}(\beta 1, \beta 2; v) = \sum_{s=-2}^{1} V_{i+r, j+s}\left( \sum_{h=0}^{3} c_{hs}(\beta 1, \beta 2)v^h \right) .$$

Rearranging,

$$Q_{ij}(u, v) = \sum_{g=0}^{3}\left( \sum_{r=-2}^{1} c_{gr}(\beta 1, \beta 2)W_{i+r, j}(\beta 1, \beta 2; v) \right) u^g \qquad (13.3)$$

where

$$W_{i+r, j}(\beta 1, \beta 2; v) = \sum_{h=0}^{3}\left( \sum_{s=-2}^{1} V_{i+r, j+s}c_{hs}(\beta 1, \beta 2) \right) v^h .$$

Denoting the inner summations by $S_{ijg}(\beta 1, \beta 2; v)$ and $T_{i+r, j, h}(\beta 1, \beta 2)$,

$$S_{ijg}(\beta 1, \beta 2; v) = \sum_{r=-2}^{1} c_{gr}(\beta 1, \beta 2)W_{i+r, j}(\beta 1, \beta 2; v)$$

and                                                                                                     (13.4)

$$T_{i+r, j, h}(\beta 1, \beta 2) = \sum_{s=-2}^{1} V_{i+r, j+s}c_{hs}(\beta 1, \beta 2) .$$

Then

$$Q_{ij}(u, v) = \sum_{g=0}^{3} S_{ijg}(\beta 1, \beta 2; v)u^g \qquad (13.5)$$

where

$$W_{i+r, j}(\beta 1, \beta 2; v) = \sum_{h=0}^{3} T_{i+r, j, h}(\beta 1, \beta 2)v^h .$$

Rewriting equation (13.5) to reduce the number of multiplications yields the following expressions for $\mathbf{Q}_{ij}(u, v)$;

$$\mathbf{Q}_{ij}(u, v) = \mathbf{S}_{ij0}(\beta 1, \beta 2; v) + u(\mathbf{S}_{ij1}(\beta 1, \beta 2; v)$$
$$+ u(\mathbf{S}_{ij2}(\beta 1, \beta 2; v) + u\mathbf{S}_{ij3}(\beta 1, \beta 2; v))) \tag{13.6}$$

where

$$\mathbf{S}_{ijg}(\beta 1, \beta 2; v) = \sum_{r=-2}^{1} c_{gr}(\beta 1, \beta 2)\mathbf{W}_{i+r, j}(\beta 1, \beta 2; v) \qquad \text{for } g = 0, 1, 2\ 3$$

and where

$$\mathbf{W}_{i+r, j}(\beta 1, \beta 2; v) = \mathbf{T}_{i+r, j, 0}(\beta 1, \beta 2) + v(\mathbf{T}_{i+r, j, 1}(\beta 1, \beta 2) + v(\mathbf{T}_{i+r, j, 2}(\beta 1, \beta 2)$$
$$+ v\mathbf{T}_{i+r, j, 3}(\beta 1, \beta 2)))$$

and where

$$\mathbf{T}_{i+r, j, h}(\beta 1, \beta 2) = \sum_{s=-2}^{1} \mathbf{V}_{i+r, j+s}c_{hs}(\beta 1, \beta 2) \qquad \text{for } h = 0, 1, 2, 3 \ .$$

Using these equations, the following algorithm computes a surface composed of $m$ by $n$ patches where the $(i, j)^{\text{th}}$ patch is evaluated at $p_i + 1$ by $q_j + 1$ values of the domain parameter:

```
compute_d (beta1, beta2, d);
compute_delta (d, delta);
compute_all_c (delta, d, c);
for i := 1 to m do
  for j := 1 to n do
  begin
    for r := −2 to 1 do
    begin
      for h := 0 to 3 do
      begin
        T_{i+r,j,h} := V_{i+r,j−2} * c_{h, −2};
        for s := −1 to 1 do
          T_{i+r,j,h} := T_{i+r,j,h} + V_{i+r,j+s} * c_{hs}
      end;
      for each v in {v_{j_k}|k = 0, 1, ..., q_j} do
      begin
        W_{i+r,j} := T_{i+r,j,3};
        for h := 2 downto 0 do
          W_{i+r,j} := T_{i+r,j,h} + v * W_{i+r,j}
      end
    end;
```

**for each** $v$ **in** $\{v_{j_k}|k = 0, 1, \ldots, q_j\}$ **do**
**begin**
   **for** $g := 0$ **to** $3$ **do**
   **begin**
      $\mathbf{S}_{ijg} := c_{g,-2} * \mathbf{W}_{i-2,j}$;
      **for** $r := -1$ **to** $1$ **do**
         $\mathbf{S}_{ijg} := \mathbf{S}_{ijg} + c_{gr} * \mathbf{W}_{i+r,j}$
   **end**;
   **for each** $u$ **in** $\{u_{i_k}|k = 0, 1, \ldots, p_i\}$ **do**
   **begin**
      $\mathbf{Q}_{ij} := \mathbf{S}_{ij3}$;
      **for** $g := 2$ **downto** $0$ **do** $\mathbf{Q}_{ij} := \mathbf{S}_{ijg} + u * \mathbf{Q}_{ij}$
   **end**
   **end**
**end**;

This algorithm requires $8 + \sum\limits_{i=1}^{m} \sum\limits_{j=1}^{n} 3\{4[4[4(1 + 3) + (q_j + 1)3] + (q_j + 1) \cdot$
$[4(1 + 3) + (p_i + 1)3]\}$ multiplications, $12 + 2 + \sum\limits_{i=1}^{m} \sum\limits_{j=1}^{n} 3\{4[4[4(3) + (q_j + 1)3] +$
$(q_j + 1)[4(3) + (p_i + 1)3]\}$ additions/subtractions, and 9 divisions. Using the nota-
tion defined in equation (8.1), and rearranging, these computational requirements
become $8 + 3mn[4(3q + 19) + (3p + 19)(q + 1)]$ multiplications, $14 + 9mn[4(q + 5)$
$+ (q + 5)(q + 1)$ additions/subtractions, and 9 divisions.
    Note that this algorithm computes the general surface evaluated at a different
set of values of the domain parameters on each patch, with no loss of efficiency.
Analogous to the curve evaluation algorithms with uniform shape parameters, it
can be seen that evaluation method I is more efficient if there are many patches or
if evaluation is sparse, and evaluation method II is preferable if there are only a few
patches or if evaluation is dense.

## 13.3 Perturbation Due to the Movement of a Control Vertex

Analogous to curve perturbation, the modification of an already-existing surface
does not require the recomputation of the entire surface; rather, the modification
of an existing surface can be accomplished much more efficiently by exploiting
various properties of the Beta-spline representation.
    Consider the consequences to an existing surface when the position of one
control vertex is modified. Since a single control vertex affects only sixteen surface
patches, the consequences of moving one vertex are confined to sixteen patches, and
hence the movement of a control vertex requires the re-evaluation of only sixteen
patches.
    Furthermore, even these sixteen patches do not need to be completely recom-
puted. Although each of these patches is controlled by sixteen vertices, only one

vertex has changed position. Therefore, the change in each of these patches is due only to the movement of one control vertex. Recalling the mathematical formulation for the $(i, j)^{th}$ surface patch given in equation (12.4), the change in this patch $Q_{ij}^A(u, v)$ can be written as

$$Q_{ij}^A(u, v) = \sum_{r=-2}^{1} [b_r(\beta 1, \beta 2; u) \sum_{s=-2}^{1} V_{i+r, j+s}^A b_s(\beta 1, \beta 2; v)] \tag{13.7}$$

where

$V_{ij}^A$ is the change in position of the control vertex $V_{ij}$ .

Assuming that the modified control vertex is the only one that has moved,

$$V_{ij}^A = 0 \qquad\qquad \text{for } i \neq \hat{i} \text{ or } j \neq \hat{j} \tag{13.8}$$

where

$V_{\hat{i}\hat{j}}$ denotes the modified vertex.

Hence, equation (13.7) contains only one nonzero term; thus, it reduces to

$$Q_{ij}^A(u, v) = b_r(\beta 1, \beta 2; u) V_{i+r, j+s}^A b_s(\beta 1, \beta 2; v) \tag{13.9}$$

where

$i + r = \hat{i}$ and $j + s = \hat{j}$ .

Rewriting equation (13.9) as

$$Q_{\hat{i}-r, \hat{j}-s}^A(u, v) = b_r(\beta 1, \beta 2; u) V_{\hat{i}\hat{j}}^A b_s(\beta 1, \beta 2; v) \tag{13.10}$$

$$\text{for } r = -2, -1, 0, 1 \quad \text{and} \quad s = -2, -1, 0, 1 \ ,$$

it is easily seen that the sixteen affected surface patches are

$$Q_{ij}(u, v) \tag{13.11}$$

where

$$i = \hat{i} - r \qquad\qquad \text{for } r = -2, -1, 0, 1$$

and

$$j = \hat{j} - s \qquad\qquad \text{for } s = -2, -1, 0, 1 \ .$$

Hence the movement of control vertex $V_{\hat{i}\hat{j}}$ perturbs the patches $Q_{ij}(u, v)$ by

$$Q_{ij}^A(u, v) = b_{\hat{i}-i}(\beta 1, \beta 2; u) V_{\hat{i}\hat{j}}^A b_{\hat{j}-j}(\beta 1, \beta 2; v) \tag{13.12}$$

where

$i$ and $j$ take on the values specified by equation (13.11).

Since equation (13.12) represents the change in the surface patch $Q_{ij}(u, v)$, the new patch can be computed by incrementing the old patch by this change:

$$Q_{ij}^{new}(u, v) = Q_{ij}^{old}(u, v) + b_{\hat{i}-i}(\beta 1, \beta 2; u) V_{\hat{i}\hat{j}}^A b_{\hat{j}-j}(\beta 1, \beta 2; v) \ . \tag{13.13}$$

The sixteen perturbed surface patches resulting from modifying the position of the control vertex $V_{\hat{i}\hat{j}}$ can be computed by the following algorithm:

**for** $j := \hat{j} - 1$ **to** $\hat{j} + 2$ **do**
  **for each** $v$ **in** $\{v_{j_k} | k = 0, 1, \ldots, q_j\}$ **do**
  **begin**
    compute_current_b $(\hat{j} - j, c, v, b)$;
    $\mathbf{W} := \mathbf{V}_{ij}^{A} * b_{\hat{j}-j}$;
    **for** $i := \hat{i} - 1$ **to** $\hat{i} + 2$ **do**
      **for each** $u$ **in** $\{u_{i_k} | k = 0, 1, \ldots, p_i\}$ **do**
      **begin**
        compute_current_b $(\hat{i} - i, c, u, b)$;
        $\mathbf{Q}_{ij}^{new}(u, v) := \mathbf{Q}_{ij}^{old}(u, v) + b_{\hat{i}-i} * \mathbf{W}$
      **end**
  **end**;

where the compute_current_b algorithm was given in Section 8.4. For comparison
purposes, assume that the sixteen affected surface patches were computed at the
same set of $p + 1$ by $q + 1$ values of the domain parameters. Then the computational requirements of this algorithm could be expressed as $24(q + 1)(4p + 5)$
multiplications and $(q + 1)(76p + 83)$ additions/subtractions. However, the computational requirements can be reduced to $21(q + 1)(4p + 5)$ multiplications and
$3(q + 1)(28p + 31)$ additions/subtractions and the introduction of the compute_
current_b algorithm can be avoided as shown in the following algorithm:

**for each** $v$ **in** $\{v_k | k = 0, 1, \ldots, q\}$ **do**
**begin**
  compute_b $(c, v, b)$;
  **for** $j := \hat{j} - 1$ **to** $\hat{j} + 2$ **do**
  **begin**
    $\mathbf{W} := \mathbf{V}_{ij}^{A} * b_{\hat{j}-j}$;
    **for each** $u$ **in** $\{u_k | k = 0, 1, \ldots, p\}$ **do**
    **begin**
      compute_b $(c, u, b)$;
      **for** $i := \hat{i} - 1$ **to** $\hat{i} + 2$ **do**
        $\mathbf{Q}_{ij}^{new}(u, v) := \mathbf{Q}_{ij}^{old}(u, v) + b_{\hat{i}-i} * \mathbf{W}$
    **end**
  **end**
**end**;

In addition, if the basis functions were precomputed for this set of $p + 1$ by $q + 1$
values of the domain parameters and stored in a table having $4(p + 1 + q + 1)$
storage locations, then a saving of $(q + 1)(9) + (p + 1)(q + 1)(4)(9)$ multiplications
and $(q + 1)(9) + (p + 1)(q + 1)(4)(9)$ additions/subtractions could be realized. The
following algorithm can be used to compute the sixteen perturbed surface patches

at these parametric values:

> **for each** $v$ **in** $\{v_k | k = 0, 1, \ldots, q\}$ **do**
>     **for** $j := \hat{j} - 1$ **to** $\hat{j} + 2$ **do**
>     **begin**
>         $\mathbf{W} := \mathbf{V}_{ij}^{A} * b_{j-j};$
>         **for each** $u$ **in** $\{u_k | k = 0, 1, \ldots, p\}$ **do**
>             **for** $i := \hat{i} - 1$ **to** $\hat{i} + 2$ **do**
>                 $\mathbf{Q}_{ij}^{new}(u, v) := \mathbf{Q}_{ij}^{old}(u, v) + b_{\hat{i}-i} * \mathbf{W}$
>     **end**;

This algorithm requires a total of $12(q + 1)(4q + 5)$ multiplications and $48(q + 1)(p + 1)$ additions/subtractions.

As with curve perturbation, this algorithm is based on the assumption that no more than one control vertex has been moved. To perturb the surface by moving more than one vertex, they must be moved one at a time, and the algorithm must be performed after each such change. The algorithm is not valid if several vertices are modified simultaneously. And again, round-off error could be problematic.

Note again that these computational requirements are not symmetrical in $p$ and $q$. If an application requires disparate values, care should be taken to write the algorithm so as to optimize the computational requirements relative to these values.

# 14 Generalizing to Continuous Shape Parameters for Surfaces

In the preceding section, the evaluation and perturbation of a Beta-spline surface has been investigated under the assumption that $\beta 1$ and $\beta 2$ are *uniform* shape parameters. Analogous to the Beta-spline curve, they will now be generalized to be *continuous* shape parameters, each varying continuously along the surface. The continuous analogues of $\beta 1$ and $\beta 2$ will be denoted $\beta 1_{ij}(u, v)$ and $\beta 2_{ij}(u, v)$, respectively, and describe the value of each shape parameter at the point $Q_{ij}(u, v)$, $i = 1, 2, \ldots, m$ and $j = 1, 2, \ldots, n$.

It is necessary that each shape parameter have a unique value at each border curve between surface patches (Figure 14.1); specifically,

$$\beta 1_{ij}(u, 1) = \beta 1_{i,j+1}(u, 0)$$

$$\beta 1_{i,j+1}(1, v) = \beta 1_{i+1,j+1}(0, v)$$

$$\beta 1_{i+1,j+1}(u, 0) = \beta 1_{i+1,j}(u, 1)$$

$$\beta 1_{i+1,j}(0, v) = \beta 1_{ij}(1, v)$$

$$\text{for } i = 1, 2, \ldots, m - 1 \quad \text{and} \quad j = 1, 2, \ldots, n - 1$$

and (14.1)

$$\beta 2_{ij}(u, 1) = \beta 2_{i,j+1}(u, 0)$$

$$\beta 2_{i,j+1}(1, v) = \beta 2_{i+1,j+1}(0, v)$$

$$\beta 2_{i+1,j+1}(u, 0) = \beta 2_{i+1,j}(u, 1)$$

$$\beta 2_{i+1,j}(0, v) = \beta 2_{ij}(1, v)$$

$$\text{for } i = 1, 2, \ldots, m - 1 \quad \text{and} \quad j = 1, 2, \ldots, n - 1 \ .$$

The actual value at a corner point can be specified by the user as a *discrete* shape parameter; these discrete analogues will be denoted $\alpha 1_{ij}$ and $\alpha 2_{ij}$. The following conditions must be satisfied by $\beta 1_{ij}(u, v)$ and analogous ones hold for $\beta 2_{ij}(u, v)$:

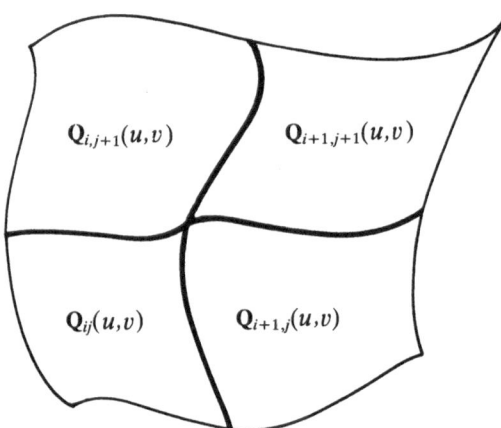

**Fig. 14.1.** Shape parameters must have a unique value at each boundary.

$$\alpha 1_{ij} = \beta 1_{ij}(1, 1)$$
$$= \beta 1_{i,j+1}(1, 0)$$
$$= \beta 1_{i+1,j+1}(0, 0)$$
$$= \beta 1_{i+1,j}(0, 1) \quad \text{for } i = 1, 2, \ldots, m - 1 \quad \text{and} \quad j = 1, 2, \ldots, n - 1$$
$$\alpha 1_{0j} = \beta 1_{1j}(0, 1)$$
$$= \beta 1_{1,j+1}(0, 0) \qquad\qquad\qquad\qquad \text{for } j = 1, 2, \ldots, n - 1$$
$$\alpha 1_{in} = \beta 1_{in}(1, 1)$$
$$= \beta 1_{i+1,n}(0, 1) \qquad\qquad\qquad\qquad \text{for } i = 1, 2, \ldots, m - 1 \quad (14.2)$$
$$\alpha 1_{mj} = \beta 1_{mj}(1, 1)$$
$$= \beta 1_{m,j+1}(1, 0) \qquad\qquad\qquad\qquad \text{for } j = 1, 2, \ldots, n - 1$$
$$\alpha 1_{i0} = \beta 1_{i1}(1, 0)$$
$$= \beta 1_{i+1,1}(0, 0) \qquad\qquad\qquad\qquad \text{for } i = 1, 2, \ldots, m - 1$$
$$\alpha 1_{00} = \beta 1_{11}(0, 0)$$
$$\alpha 1_{0n} = \beta 1_{1n}(0, 1)$$
$$\alpha 1_{mn} = \beta 1_{mn}(1, 1)$$
$$\alpha 1_{m0} = \beta 1_{m1}(1, 0)$$

Analogous to equation (6.3), the following $\beta 1_{ij}(u, v)$ and $\beta 2_{ij}(u, v)$ functions are a solution to these conditions:

$$\beta 1_{ij}(u, v) = s\, t\, \alpha 1_{ij} + (1 - s)\, t\, \alpha 1_{i-1,j} + s\, (1 - t)\, \alpha 1_{i,j-1}$$
$$+ (1 - s)\, (1 - t)\, \alpha 1_{i-1,j-1}$$
$$\beta 2_{ij}(u, v) = s\, t\, \alpha 2_{ij} + (1 - s)\, t\, \alpha 2_{i-1,j} + s\, (1 - t)\, \alpha 2_{i,j-1} \qquad (14.3)$$
$$+ (1 - s)\, (1 - t)\, \alpha 2_{i-1,j-1}$$

where

$$s = 10u^3 - 15u^4 + 6u^5$$

and

$$t = 10v^3 - 15v^4 + 6v^5$$

$$\text{for } i = 1, 2, \ldots, m \quad \text{and} \quad j = 1, 2, \ldots, n \ .$$

In addition, it is necessary to extend the notation

$$\delta = 2 \, \beta1^3 + 4 \, \beta1^2 + 4\beta1 + \beta2 + 2$$

to

$$\delta_{ij}(u, v) = 2 \, \beta1_{ij}^3(u, v) + 4 \, \beta1_{ij}^2(u, v) + 4 \, \beta1_{ij}(u, v) + \beta2_{ij}(u, v) + 2 \tag{14.4}$$

$$\text{for } i = 1, 2, \ldots, m \quad \text{and} \quad j = 1, 2, \ldots, n \ .$$

Finally, the discrete analogue to $\delta_{ij}(u, v)$ is

$$\gamma_{00} = \delta_{11}(0, 0)$$

$$\gamma_{0j} = \delta_{1j}(0, 1) \qquad\qquad\qquad \text{for } j = 1, 2, \ldots, n$$

$$\gamma_{i0} = \delta_{i1}(1, 0) \qquad\qquad\qquad \text{for } i = 1, 2, \ldots, m \tag{14.5}$$

$$\gamma_{ij} = \delta_{ij}(1, 1) \qquad\qquad \text{for } i = 1, 2, \ldots, m \quad \text{and} \quad j = 1, 2, \ldots, n \ .$$

# 15 Surface Evaluation and Perturbation with Continuous Shape Parameters

## 15.1 Evaluation Method I

An important observation is that $\beta 1_{ij}(u, v)$ and $\beta 2_{ij}(u, v)$ (equation (14.3)) can each be written as a pair of equations of similar form; specifically,

$$\beta 1_{ij}(u, v) = s\, \tau 1_{ij}(t) + (1 - s)\, \tau 1_{i-1, j}(t)$$

where

$$\tau 1_{ij}(t) = t\, \alpha 1_{ij} + (1 - t)\, \alpha 1_{i, j-1}$$

and                                                                                                        (15.1)

$$\beta 2_{ij}(u, v) = s\, \tau 2_{ij}(t) + (1 - s)\, \tau 2_{i-1, j}(t)$$

where

$$\tau 2_{ij}(t) = t\, \alpha 2_{ij} + (1 - t)\, \alpha 2_{i, j-1}$$

$$\text{for } i = 1, 2, \ldots, m \quad \text{and} \quad j = 1, 2, \ldots, n$$

where

$s$ and $t$ were defined in equation (14.3).

Generalizing evaluation method I described in Section 13.1 to *continuous* shape parameters, the following algorithm computes a general Beta-spline surface composed of $m$ by $n$ patches where the $(i, j)^{\text{th}}$ patch is evaluated at $p_i + 1$ by $q_j + 1$ values of the domain parameter:

```
for j := 1 to n do
begin
    jump_tau1 := alpha1_oj - alpha1_o,j-1;
    jump_tau2 := alpha2_oj - alpha2_o,-1;
    for i := 1 to m do
    begin
        jump_sigma1 := jump_tau1;
```

```
jump_tau1 := alpha1_{ij} - alpha1_{i,j-1};
jump_sigma2 := jump_tau2;
jump_tau2 := alpha2_{ij} - alpha2_{i,j-1};
for each v in {v_{j_k}|k = 0, 1, ..., q_j} do
begin
    t := (10 + (6 * v - 15) * v) * v * v * v;
    sigma1 := alpha1_{i-1,j-1} + t * jump_sigma1;
    tau1 := alpha1_{i,j-1} + t * jump_tau1;
    jump_beta1 := tau1 - sigma1;
    sigma2 := alpha2_{i-1,j-1} + t * jump_sigma2;
    tau2 := alpha2_{i,j-1} + t * jump_tau2;
    jump_beta2 := tau2 - sigma2;
    for each u in {u_{i_k}|k = 0, 1, ..., p_i} do
    begin
        s := (10 + (6 * u - 15) * u) * u * u * u;
        beta1 := sigma1 + s * jump_beta1;
        beta2 := sigma2 + s * jump_beta2;
        compute_d (beta1, beta2, d);
        compute_delta (d, delta);
        delta2 := delta * delta;
        compute_a (d, delta, v, a);
        for r := -2 to 1 do
        begin
            compute_W (i, r, j, a, V, W);
            compute_a (d, delta, u, a);
            compute_Q (i, j, a, delta2, W, Q)
        end
    end
end
end;
where
procedure compute_Q (i; j; a; delta2; W; Q);
begin (* compute_Q *)
    Q_{ij} := a_{-2} * W_{i-2,j};
    for r := -1 to 1 do Q_{ij} := Q_{ij} + a_r * W_{i+r,j};
    Q_{ij} := Q_{ij}/delta2
end (* compute_Q *);
```

The compute_$Q$ algorithm requires $3(1 + 3) = 12$ multiplications, $3(3) = 9$ additions/subtractions, and $3$ divisions. The complete algorithm then requires a total of $\sum\limits_{i=1}^{m} \sum\limits_{j=1}^{n} [(q_j + 1)\{5 + 4 + (p_i + 1)(5 + 2 + 7 + 1 + 9 + 4(12) + 9 + 12)\}]$

multiplications, $\sum\limits_{j=1}^{n} (2 + \sum\limits_{i=1}^{m} [2 + (q_j + 1)\{2 + 6 + (p_i + 1)(2 + 2 + 12 + 2 + 9$

$+ 4(9) + 9 + 9)\}])$ additions/subtractions, and $\sum\limits_{i=1}^{m} \sum\limits_{j=1}^{n} 3(p_i + 1)(q_j + 1)$ divisions.

Using the notation defined in equation (8.1), and rearranging, these computational requirements become $3mn(31p + 34)(q + 1)$ multiplications, $n(2 + m[2 + (81p + 89)(q + 1)])$ additions/subtractions, and $3mn(p + 1)(q + 1)$ divisions. It should be noted that this algorithm naturally computes the general surface evaluated at a different set of values of the domain parameter on each patch, with no loss of efficiency.

The algorithm given in Section 13.2 was generalized to continuous shape parameters, but it was found that the result was very complex to program and was less efficient than the algorithm given above.

## 15.2 Perturbation Due to the Movement of a Control Vertex

Similar to the perturbation of a Beta-spline surface with uniform shape parameters explained in Section 13.3, an already-existing surface with continuous shape parameters can be modified in a manner that is more efficient than a complete recomputation.

The following algorithm perturbs a general surface where the $(i, j)^{th}$ surface patch is computed at $p_i + 1$ by $q_j + 1$ values of the domain parameters:

```
for j := ĵ − 1 to ĵ + 2 do
begin
    jump_tau1 := alpha1_{i−2,j} − alpha1_{i−2,j−1};
    jump_tau2 := alpha2_{i−2,j} − alpha2_{i−2,j−1};
    for i := î − 1 to î + 2 do
    begin
        jump_sigma1 := jump_tau1;
        jump_tau1 := alpha1_{ij} − alpha1_{i,j−1};
        jump_sigma2 := jump_tau2;
        jump_tau2 := alpha2_{ij} − alpha2_{i,j−1};
        for each v in {v_{j_k}|k = 0, 1, ..., q_j} do
        begin
            t := (10 + (6 * v − 15) * v) * v * v * v;
            sigma1 := alpha1_{i−1,j−1} + t * jump_sigma1;
            tau1 := alpha1_{i,j−1} + t * jump_tau1;
            jump_beta1 := tau1 − sigma1;
```

$sigma2 := alpha2_{i-1,j-1} + t*jump\_sigma2;$

$tau2 := alpha2_{i,j-1} + t*jump\_tau2;$

$jump\_beta2 := tau2 - sigma2;$

**for each** $u$ **in** $\{u_{i_k}|k = 0, 1, \ldots, p_i\}$ **do**

**begin**

    $s := (10 + (6*u - 15)*u)*u*u*u;$

    $beta1 := sigma1 + s*jump\_beta1;$

    $beta2 := sigma2 + s*jump\_beta2;$

    compute_d $(beta1, beta2, d);$

    compute_delta $(d, delta);$

    compute_current_a $(\hat{j} - j, d, v, a);$

    $b_{\hat{j}-j} := a_{\hat{j}-j}/delta;$

    $\mathbf{W} := \mathbf{V}_{ij}^{A} * b_{\hat{j}-j};$

    compute_current_a $(\hat{i} - i, d, u, a);$

    $b_{\hat{i}-i} := a_{\hat{i}-i}/delta;$

    $\mathbf{Q}_{ij}^{new}(u, v) := \mathbf{Q}_{ij}^{old}(u, v) + b_{\hat{i}-i} * \mathbf{W}$

    **end**

   **end**

  **end**

 **end;**

where the compute_current_a algorithm was given in Section 10.3. For comparison purposes, assume that the sixteen affected surface patches were computed at the same set of $p + 1$ by $q + 1$ values of the domain parameters. Then the computational requirements of this algorithm could be expressed as $16(q + 1)(26p + 35)$ multiplications, $40 + 8(q + 1)(53p + 69)$ additions/subtractions, and $32(q + 1)(p + 1)$ divisions.

## 15.3  Perturbation Due to the Modification of Shape Parameters

A Beta-spline surface with continuous shape parameters can also be locally modified by altering the values of the shape parameters at a corner point. Note that the continuous shape parameters for the $(i, j)^{th}$ surface patch, $\beta1_{ij}(u, v)$ and $\beta2_{ij}(u, v)$, are functions of $\alpha1_{i-1,j-1}, \alpha1_{i-1,j}, \alpha1_{i,j}$ and $\alpha1_{i,j-1}$, and of $\alpha2_{i-1,j-1}, \alpha2_{i-1,j}, \alpha2_{i,j}$ and $\alpha2_{i,j-1}$, respectively. Denoting the modified shape parameters as $\alpha1_{ij}$ and $\alpha2_{ij}$, the affected surface patches are $\mathbf{Q}_{ij}(u, v)$, $\mathbf{Q}_{i,j+1}(u, v)$, $\mathbf{Q}_{i+1,j+1}(u, v)$, and $\mathbf{Q}_{i+1,j}(u, v)$. Thus, changing the values of the shape parameters at one corner point requires the re-evaluation of only four surface patches.

# 16 Classification and Analysis of Beta-spline Surface Boundary Conditions

## 16.1 Classification

The $(m + 1) \times (n + 1)$ control graph described in Section 7.1 (Figure 2.2) contains the set of control vertices

$$\mathbf{V} = \{\mathbf{V}_{ij} | i = 0, 1, \ldots, m, \ j = 0, 1, \ldots, n\} \ .$$

Using the Beta-spline surface formulation (equation (12.1)), it can be seen that these vertices naturally define the interior patches

$$\mathbf{Q}_{ij}(u, v) \qquad\qquad i = 2, 3, \ldots, m - 1; \ j = 2, 3, \ldots, n - 1$$

(Figure 16.1). It is desirable to have additional patches around the periphery which are more dominated by the boundary vertices. Analogous to the Beta-spline curve formulation, such additional patches cannot be defined simply by evaluating the Beta-spline surface formulation (equation (12.1)) in the usual manner because this would reference nonexistent vertices. This problem can be circumvented using two different types of boundary condition techniques, *multiple vertices* and *phantom vertices*, which will now be described.

## 16.2 Analysis of Multiple Vertices Boundary Conditions

### 16.2.1 Double Vertices

With this technique, additional surface patches are defined around the periphery of the interior patches which were naturally defined by the control graph, by repeating boundary vertices in the Beta-spline surface formulation. The interior patches are then surrounded by the additional patches (see Figure 16.2):

$$\mathbf{Q}_{1j}(u, v), \quad j = 1, 2, \ldots, n - 1;$$

$$\mathbf{Q}_{in}(u, v), \quad i = 1, 2, \ldots, m - 1;$$

$$\mathbf{Q}_{mj}(u, v), \quad j = 2, 3, \ldots, n; \quad \text{and}$$

$$\mathbf{Q}_{i1}(u, v), \quad i = 2, 3, \ldots, m \ .$$

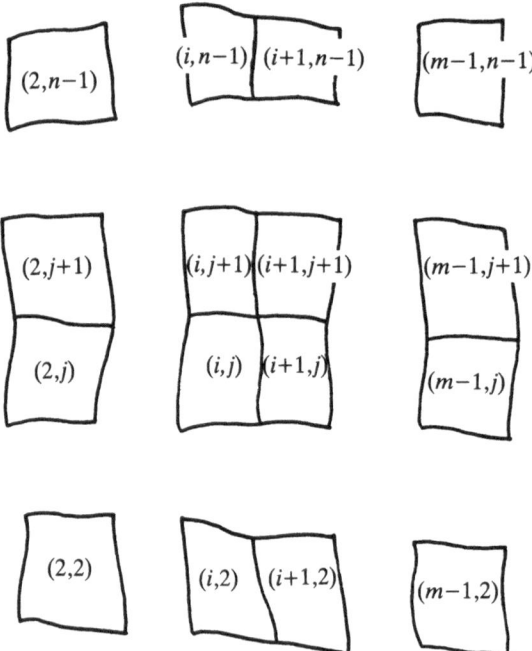

**Fig. 16.1.**  Interior patches naturally defined by the control graph.

The additional patches are defined by evaluating the usual Beta-spline surface formulation (equation (12.1)) except that whenever a nonexistent vertex is referenced, the "nearest" boundary vertex is used instead. Specifically, let $V_{ij}$ be the referenced vertex. Then the selection of the appropriate vertex is accomplished as follows:

if $i < 0$ **then use** $V_{0j}$

    **else if** $i > m$ **then use** $V_{mj}$;

if $j < 0$ **then use** $V_{i0}$

    **else if** $j > n$ **then use** $V_{in}$;

Note that an out-of-range value of the subscript $i$ does not preclude an out-of-range value of the subscript $j$.

This technique yields a Beta-spline surface composed of $m$ by $n$ patches. Since the control graph has $m + 1$ by $n + 1$ vertices which naturally define an array of $m$ by $n$ regions, this is a convenient arrangement of patches.

To analyze the curvature at the endpoint along a boundary curve between adjacent boundary patches requires the evaluation of the first and second derivative vectors there. This can be accomplished in a manner analogous to that which was described in Chapter 11, with the result that the second derivative vector is a multiple of the first derivative vector which is, in general, nonzero. This satisfies Condition 1 of Section 4.5.4; thus, the curvature is, in general, zero at the endpoint along a boundary curve.

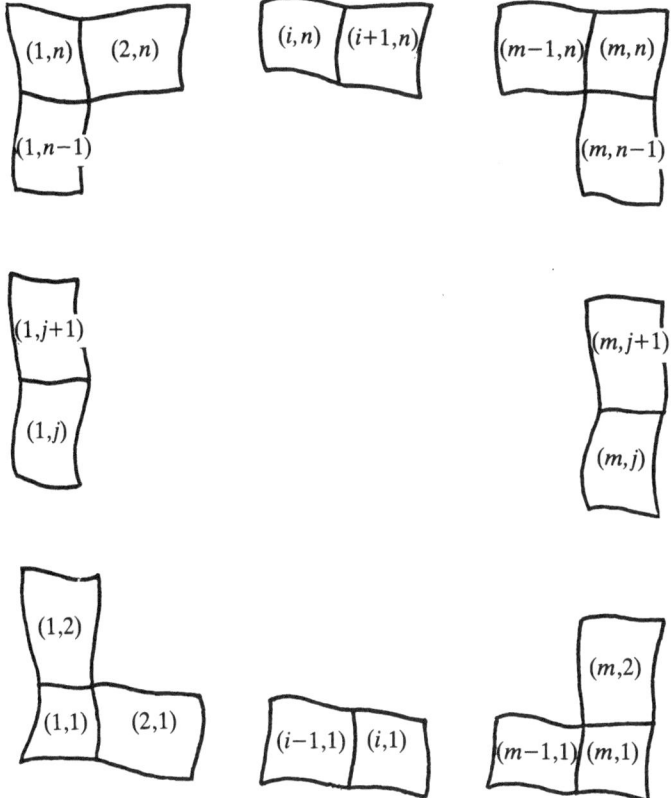

**Fig. 16.2.** Additional patches—double vertices boundary condition.

Analogous to the double vertices end conditions for curves, the surface patches defined by this technique are, in general, smaller than those defined by the unmodified Beta-spline surface formulation. This characteristic is due to the fact that the vertices that define such a patch are not all distinct; thus, they are more densely clustered than those which define a regular patch.

Finally, it is of interest to consider the value of the *twist vector*; that is, the parametric mixed partial derivative vector, at each of the four corner points of the entire surface. For notational convenience, the twist vector at $(c, d)$ will be denoted $Q^{(1,1)}(u, v)$; that is

$$Q^{(1,1)}(c, d) = \frac{\partial^2}{\partial u \partial v} Q(u, v)\bigg|_{u=c, v=d} \quad .$$

Differentiating equation (12.1) with respect to $u$ and $v$, and then evaluating at the appropriate parametric values by substituting the values of the first derivative of each basis function as given in Section 7.4, results in the following expressions for these values of the twist vector:

$$Q_{11}^{(1,1)}(0, 0) = \frac{36}{\gamma_{00}^2}[\mathbf{V}_{00} - \mathbf{V}_{01} + \mathbf{V}_{11} - \mathbf{V}_{10}]$$

$$Q_{1n}^{(1,1)}(0, 1) = \frac{36}{\gamma_{0n}^2}[\mathbf{V}_{0,n-1} - \mathbf{V}_{0n} + \mathbf{V}_{1n} - \mathbf{V}_{1,n-1}]$$

(16.1)

$$Q_{mn}^{(1,1)}(1, 1) = \frac{36}{\gamma_{mn}^2}[\mathbf{V}_{m-1,n-1} - \mathbf{V}_{m-1,n} + \mathbf{V}_{mn} - \mathbf{V}_{m,n-1}]$$

$$Q_{m1}^{(1,1)}(1, 0) = \frac{36}{\gamma_{m0}^2}[\mathbf{V}_{m-1,0} - \mathbf{V}_{m-1,1} + \mathbf{V}_{m1} - \mathbf{V}_{m0}] \ .$$

Note that each is a scalar multiple of the constant twist vector of the corresponding *bilinear* surface whose corner points are the four vertices appearing in the expression for the twist vector, where the scale factor is $\frac{36}{\gamma_{ij}^2}$ for patch $\mathbf{Q}_{ij}(u, v)$.

### 16.2.2 Triple Vertices

This technique extends the double vertices technique by defining another set of additional surface patches around the periphery of those which were defined by the double vertices technique. This second set of additional patches is

$$\mathbf{Q}_{0j}(u, v), \qquad j = 0, 1, \ldots, n;$$

$$\mathbf{Q}_{i,n+1}(u, v), \quad i = 0, 1, \ldots, m;$$

$$\mathbf{Q}_{m+1,j}(u, v), \quad j = 1, 2, \ldots, n + 1; \quad \text{and}$$

$$\mathbf{Q}_{i0}(u, v), \qquad i = 1, 2, \ldots, m + 1$$

(Figure 16.3) and they are defined by the same algorithm as that for the patches defined by the double vertices technique.

Using this algorithm for the patch $\mathbf{Q}_{0j}(u, v)$ yields

$$\mathbf{Q}_{0j}(u, v) = (b_{-2}(u) + b_{-1}(u) + b_0(u)) \sum_{s=-2}^{1} \mathbf{V}_{0,j+s}b_s(v)$$

$$+ b_1(u) \sum_{s=-2}^{1} \mathbf{V}_{1,j+s}b_s(v) \tag{16.2}$$

and then substituting the expressions for the basis functions results in

$$\mathbf{Q}_{0j}(u, v) = \left(1 - \frac{2u^3}{\delta_{ij}(u, v)}\right) \sum_{s=-2}^{1} \mathbf{V}_{0,j+s}b_s(v)$$

$$+ \frac{2u^3}{\delta_{ij}(u, v)} \sum_{s=-2}^{1} \mathbf{V}_{1,j+s}b_s(v) \ . \tag{16.3}$$

Consider the pair of boundary curves of the patch $\mathbf{Q}_{0j}(u, v)$ obtained by holding the domain parameter $u$ constant at the extreme values 0 and 1, respectively. Each is a Beta-spline curve controlled by four of the eight control vertices defining this patch. The patch is just a *ruled* (linearly lofted) surface between these curves; that

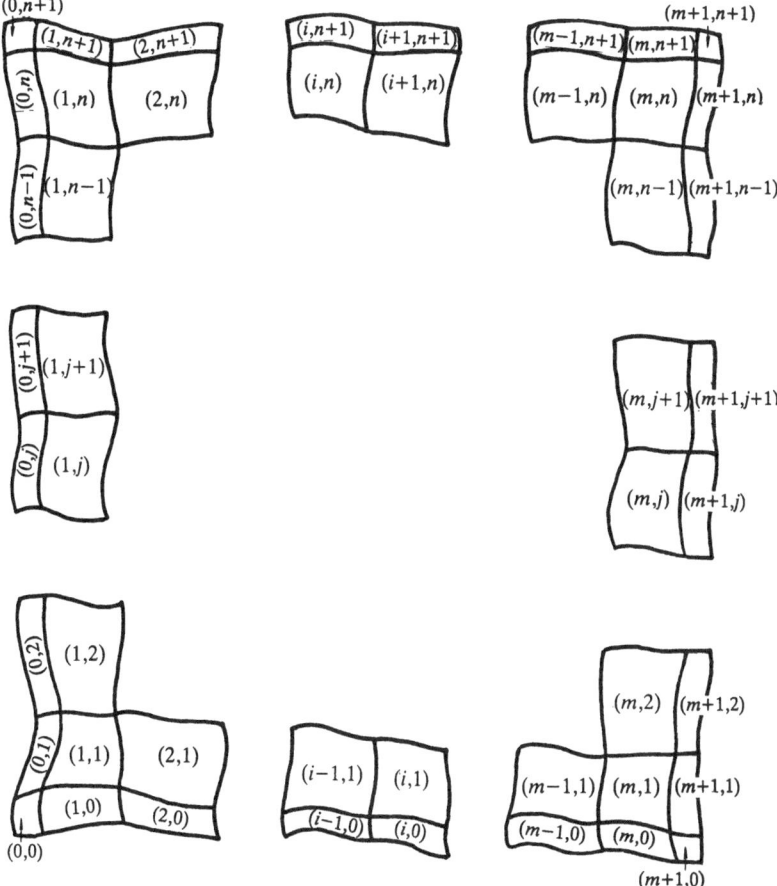

**Fig. 16.3.** Additional patches—triple vertices boundary condition.

is, for any fixed value of the domain parameter $v$, the curve traced out on this surface patch, as $u$ varies from 0 to 1, is a straight line segment between corresponding points on each of the Beta-spline boundary curves. In particular, the other pair of boundary curves (at $v = 0$ and $v = 1$) are straight line segments. A similar argument holds for all the boundary patches defined by this technique; thus, all the boundary curves between such patches are straight line segments.

Although such curve segments are straight, they are nonetheless quite short, in general. For this reason, this condition may be acceptable in some cases. Still, it is frequently undesirable when attempting to design a smooth surface and would be particularly noticeable if the surface is to be machined by a numerically controlled cutting tool.

This smaller length is just a special case of the more general tendency of these surface patches to be very small; even smaller than those defined by the double vertices technique. Again, this is because the vertices defining such a patch are not all distinct; thus, they are, in general, more densely clustered than those defining a regular patch.

Consider the four corner patches of the entire surface. These patches are given by the expressions

$$\mathbf{Q}_{00}(u, v) = [1 - f(u)][1 - f(v)]\mathbf{V}_{00} + [1 - f(u)]f(v)\mathbf{V}_{01}$$
$$+ f(u)f(v)\mathbf{V}_{11} + f(u)[1 - f(v)]\mathbf{V}_{10}$$

$$\mathbf{Q}_{0,n+1}(u, v) = [1 - f(u)]g(v)\mathbf{V}_{0,n-1} + [1 - f(u)][1 - g(v)]\mathbf{V}_{0n}$$
$$+ f(u)[1 - g(v)]\mathbf{V}_{1n} + f(u)g(v)\mathbf{V}_{1,n-1}$$

$$\mathbf{Q}_{m+1,n+1}(u, v) = g(u)g(v)\mathbf{V}_{m-1,n-1} + g(u)[1 - g(v)]\mathbf{V}_{m-1,n}$$
$$+ [1 - g(u)][1 - g(v)]\mathbf{V}_{mn} + [1 - g(u)]g(v)\mathbf{V}_{m,n-1}$$

$$\mathbf{Q}_{m+1,0}(u, v) = g(u)[1 - f(v)]\mathbf{V}_{m-1,0} + g(u)f(v)\mathbf{V}_{m-1,1}$$
$$+ [1 - g(u)]f(v)\mathbf{V}_{m1} + [1 - g(u)][1 - f(v)]\mathbf{V}_{m0}$$

(16.4)

where

$$f(t) = \frac{2t^3}{\delta_{ij}(u, v)}$$

and

$$g(t) = \frac{2 \,\beta 1_{ij}(u, v)(1 - t)^3}{\delta_{ij}(u, v)}$$

for

$$\mathbf{Q}_{ij}(u, v) \ .$$

These four corner patches are bilinear surfaces, each having straight line segments for all four of its boundary curves. These patches are, in general, very small; still smaller than the other boundary patches defined by the triple vertices technique.

The expression for the twist vector, $\mathbf{Q}^{(1, 1)}(u, v)$, on each of these four corner patches of the surface is:

$$\mathbf{Q}_{00}^{(1, 1)}(u, v) = \frac{36u^2v^2}{\gamma_{00}}[\mathbf{V}_{00} - \mathbf{V}_{01} + \mathbf{V}_{11} - \mathbf{V}_{10}]$$

$$\mathbf{Q}_{0,n+1}^{(1, 1)}(u, v) = \frac{36 \,\beta 1_{0,n+1}^3(u, v) * u^2(1 - v)^2}{\gamma_{0,n+1}}[\mathbf{V}_{0,n-1} - \mathbf{V}_{0n} + \mathbf{V}_{1n} - \mathbf{V}_{1,n-1}]$$

$$\mathbf{Q}_{m+1,n+1}^{(1, 1)}(u, v) = \frac{36 \,\beta 1_{m+1,n+1}^6(u, v) * (1 - u)^2(1 - v)^2}{\gamma_{m+1,n+1}}[\mathbf{V}_{m-1,n-1}$$

(16.5)

$$- \mathbf{V}_{m-1,n} + \mathbf{V}_{mn} - \mathbf{V}_{m,n-1}]$$

$$\mathbf{Q}_{m+1,0}^{(1, 1)}(u, v) = \frac{36 \,\beta 1_{m+1,0}^3(u, v) * (1 - u)^2v^2}{\gamma_{m+1,0}}[\mathbf{V}_{m-1,0} - \mathbf{V}_{m-1,1} + \mathbf{V}_{m1} - \mathbf{V}_{m0}] \ .$$

Note that each has a constant direction and sense which are identical to that of the constant twist vector of the corresponding *bilinear* surface whose corner points are the four vertices appearing in the expression for the twist vector.

Finally, it should be mentioned that the four corner vertices are interpolated. That is,

$$\mathbf{Q}_{00}(0, 0) = \mathbf{V}_{00}$$

$$\mathbf{Q}_{0,n+1}(0, 1) = \mathbf{V}_{0n}$$

$$\mathbf{Q}_{m+1,n+1}(1, 1) = \mathbf{V}_{mn} \tag{16.6}$$

$$\mathbf{Q}_{m+1,0}(1, 0) = \mathbf{V}_{m0} \; .$$

Note, however, that the other boundary vertices are *not*, in general, interpolated.


## 16.3  Analysis of Phantom Vertices Boundary Conditions

The surface analogue of this type of end condition for a curve creates a set of phantom vertices around the boundaries of the original control graph. These phantom vertices are used to define additional surface patches (Figure 16.4) around the patches which were naturally defined by the control graph. This is accomplished by evaluating the Beta-spline surface formulation in the same manner as for the patches defined by the original control graph. Since the additional patches are defined by the same formulation as the normal patches, they are of normal size. In addition, the complete surface consists of $m$ by $n$ patches, which is a convenient arrangement since the control graph has $m + 1$ by $n + 1$ vertices which naturally define an array of $m$ by $n$ regions. Analogous to this type of end condition for a curve, the phantom vertices are completely defined in terms of the original control vertices to satisfy some boundary condition. However, the direct specification of positions or of parametric first or second derivative vectors around the boundaries of the surface would be unwieldy. A convenient condition is to set the appropriate parametric second partial derivative vector to zero at the endpoint along each boundary curve between adjacent surface patches. The appropriate derivative is with respect to the parametric direction *across* the boundary.

Evaluating these derivatives at the appropriate parametric values, substituting the values of the second derivative of each basis function (as tabulated in Section 7.4), and setting the resulting expression to zero yields an underspecified system containing $2m + 2n + 4$ equations in the $2m + 2n + 8$ unknown phantom vertices. A solution is given below which defines each of the four corner phantom vertices in terms of other phantom vertices and expresses the other phantom vertices explicitly in terms of the original control vertices:

$$\mathbf{V}_{-1, -1} = \frac{2 \, \alpha 1_{00}^2 + \alpha 2_{00}}{2 \, \alpha 1_{00}^3} [\mathbf{V}_{0, -1} - \mathbf{V}_{1, -1}] + \mathbf{V}_{0, -1}$$

$$\mathbf{V}_{-1, n+1} = \frac{2 \, \alpha 1_{0n} + \alpha 2_{0n}}{2} [\mathbf{V}_{-1, n} - \mathbf{V}_{-1, n-1}] + \mathbf{V}_{-1, n}$$

$$\mathbf{V}_{m+1, n+1} = \frac{2 \, \alpha 1_{mn} + \alpha 2_{mn}}{2} [\mathbf{V}_{m+1, n} - \mathbf{V}_{m+1, n-1}] + \mathbf{V}_{m+1, n}$$

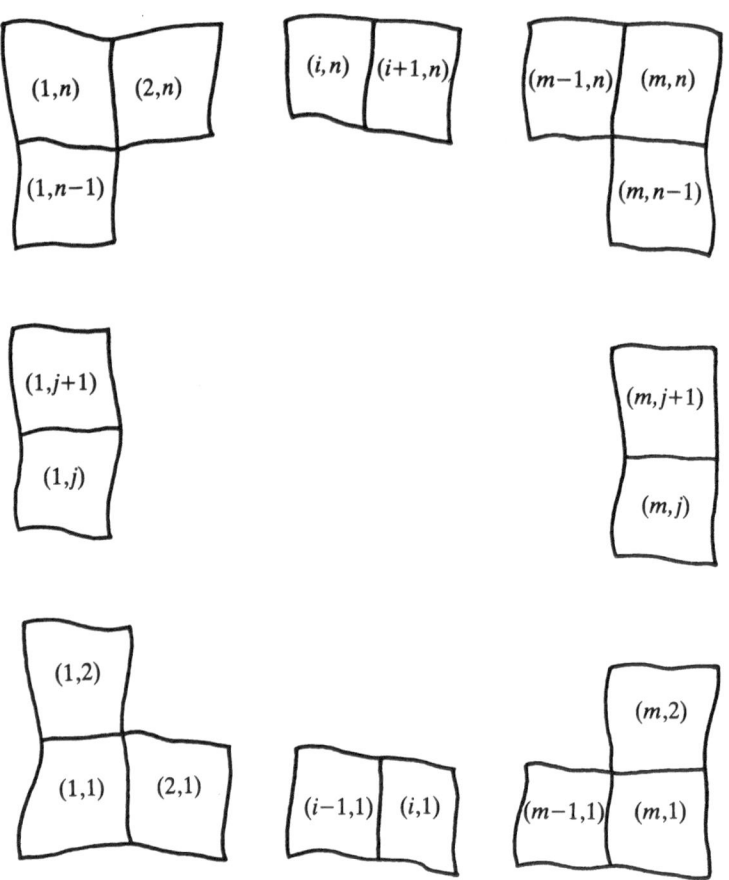

**Fig. 16.4.** Additional patches—phantom vertices boundary condition.

$$V_{m+1,\,-1} = \frac{2\,\alpha 1_{m0} + \alpha 2_{m0}}{2} \left[V_{m,\,-1} - V_{m-1,\,-1}\right] + V_{m,\,-1}$$

$$V_{-1,\,j} = \frac{2\,\alpha 1_{0j}^2 + \alpha 2_{0j}}{2\alpha 1_{0j}^3} \left[V_{0j} - V_{1j}\right] + V_{0j}$$

$$\text{for } j = 0, 1, \ldots, n$$

$$V_{i,\,n+1} = \frac{2\,\alpha 1_{in} + \alpha 2_{in}}{2} \left[V_{in} - V_{i,\,n-1}\right] + V_{in}$$

$$\text{for } i = 0, 1, \ldots, m$$

(16.7)

$$V_{m+1,\,j} = \frac{2\,\alpha 1_{mj} + \alpha 2_{mj}}{2} \left[V_{mj} - V_{m-1,\,j}\right] + V_{mj}$$

$$\text{for } j = 0, 1, \ldots, n$$

$$V_{i,\,-1} = \frac{2\,\alpha 1_{i0}^2 + \alpha 2_{i0}}{2\alpha 1_{i0}^3} \left[V_{i0} - V_{i1}\right] + V_{i0}$$

$$\text{for } i = 0, 1, \ldots, m \ .$$

Knowing that these $2m + 2n + 4$ second derivative vectors are zero, it is of interest to determine the values of the twist vector at each of the four corner points of the entire surface. This is accomplished using the values of the first derivative of each basis function as given in Section 7.4, and substituting the above expressions given in equation (16.7) for the phantom vertices. The expressions for these values of the twist vector are then

$$\mathbf{Q}_{11}^{(1,1)}(0, 0) = a_{00}(\mathbf{V}_{00} - \mathbf{V}_{01} + \mathbf{V}_{11} - \mathbf{V}_{10})$$

$$\mathbf{Q}_{1n}^{(1,1)}(0, 1) = a_{0n}(\mathbf{V}_{0,n-1} - \mathbf{V}_{0n} + \mathbf{V}_{1n} - \mathbf{V}_{1,n-1})$$

$$\mathbf{Q}_{mn}^{(1,1)}(1, 1) = a_{mn}(\mathbf{V}_{m-1,n-1} - \mathbf{V}_{m-1,n} + \mathbf{V}_{mn} - \mathbf{V}_{m,n-1})$$  (16.8)

$$\mathbf{Q}_{m1}^{(1,1)}(1, 0) = a_{m0}(\mathbf{V}_{m+1,0} - \mathbf{V}_{m+1,1} + \mathbf{V}_{m1} - \mathbf{V}_{m0})$$

where

$$a_{ij} = \frac{9}{\gamma_{ij}^2}[4\,\alpha 1_{ij}^4 + 8\,\alpha 1_{ij}^3 + 4\,\alpha 1_{ij}^2\,\alpha 2_{ij} + 4\,\alpha 1_{ij}^2 + 4\,\alpha 1_{ij}\,\alpha 2_{ij} + \alpha 2_{ij}^2]\ .$$

Each is equivalent to "$a$" times the constant bilinear twist vector, as explained in Section 16.2.1. Thus, the surface has a scaled bilinear twist vector at each of the four corner points.

Using the phantom vertices given in equation (16.7) to define the boundary patches, the first derivative vector at the endpoint along a boundary curve between such patches will, in general, be nonzero. Since the second derivative vector there is (set to) zero by this boundary condition, these derivative vectors are trivially linearly dependent. Hence, by Condition 1 of Section 4.5.4, the curvature is, in general, zero at the endpoint along a boundary curve between adjacent boundary patches.

## 16.4 Conclusion

Boundary conditions for a Beta-spline surface have been classified as *multiple vertices* and *phantom vertices*, and the multiple vertices boundary conditions considered were again *double vertices* and *triple vertices*.

The double vertices technique defines additional surface patches around the periphery of the interior patches yielding a Beta-spline surface composed of $m$ by $n$ patches. The curvature is, in general, zero at the endpoint along a boundary curve between adjacent boundary patches. Also, these boundary patches are, in general, smaller than the interior patches. The value of the twist vector at each of the four corner points of the entire surface is $1/4$ times the constant twist vector of the corresponding bilinear surface.

The triple vertices technique is an extension of the double vertices technique where another set of additional surface patches are defined around the periphery of those which were defined by the double vertices technique. The boundary curves between the patches generated by this technique are short straight line segments and each of these patches is a ruled surface between a pair of Beta-spline curves. The four corner patches are small bilinear surfaces, each having straight line segments for all four of its boundary curves. Also, the four corner vertices are interpolated, although the other boundary vertices are not.

The phantom vertices technique creates a set of phantom vertices around the boundary of the original control graph which defines additional surface patches. These patches are of normal size and the complete surface consists of $m$ by $n$ patches. These phantom vertices are defined in terms of the original control vertices such that the parametric second partial derivative vector is zero along each boundary curve between adjacent surface patches. The curvature is, in general, zero at the endpoint along a boundary curve between adjacent boundary patches. At each of the four corner points of the surface, the twist vector is equivalent to a scaled bilinear twist vector.

# 17 Geometrical Interpretation of the Shape Parameters

## 17.1 Introduction

The two inherent *shape parameters* of the Beta-spline, $\beta 1$ and $\beta 2$, provide further control of shape in addition to, and independent of, that supplied by the positional information specified by the control vertices. These shape parameters have the property that $\beta 1 = 1$ indicates continuity of the parametric first derivative vector and $\beta 1 = 1$ with $\beta 2 = 0$ indicates continuity of the parametric first and second derivative vectors.

## 17.2 $\beta 1$: Bias

Considering the middle equation in (7.1), it can be seen that the absolute value of $\beta 1$ is the ratio of magnitudes of the first parametric derivative vector at a joint. Specifically, it measures the relative magnitude of the first parametric derivative vector at the beginning of a curve segment and that at the end of the preceding curve segment.

Recall that these curve segments meet with a continuous unit tangent vector. Intuitively, $\beta 1$ measures the relative influence of this tangent direction on either side of the joint, and thus is called the *bias* at the joint. Values in excess of unity indicate that the curve will continue close to the tangent direction more on the $(i + 1)^{st}$ curve segment than on the $i^{th}$ curve segment, while values less than unity indicate the opposite situation, and $\beta 1 = 1$ means an equal influence on both sides of the joint. Figures 17.1, 17.2, and 17.3 show Beta-spline curves with uniform shape parameter values of $\beta 1 = 1$ with $\beta 2 = 0$, of $\beta 1 = 2$ with $\beta 2 = 0$, and of $\beta 1 = 0.5$ with $\beta 2 = 0$, respectively.

Consider a Beta-spline curve when both $\beta 1$ and $\beta 2$ are zero. Evaluating equation (7.11) at these values and substituting into equation (7.2) shows that

$$Q_{i+1}(0) = Q_i(1) = V_{i+1} \ . \tag{17.1}$$

This vertex interpolation is illustrated in Figure 17.4 which shows a Beta-spline curve with uniform shape parameter values of $\beta 1 = \beta 2 = 0$.

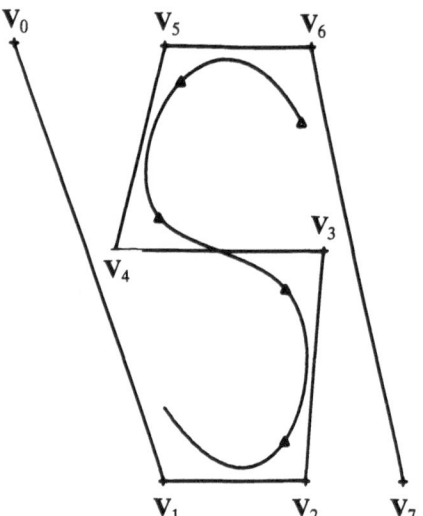

**Fig. 17.1.**   $\beta 1 = 1$ with $\beta 2 = 0$.

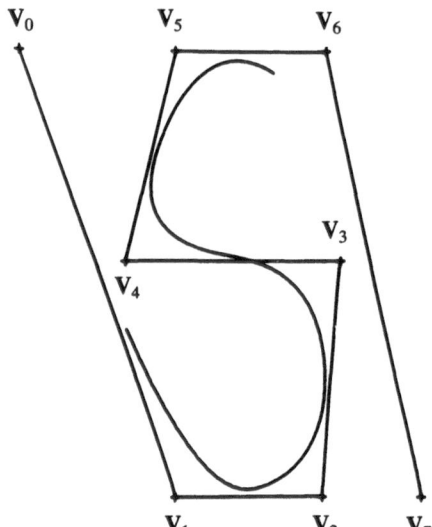

**Fig. 17.2.**   $\beta 1 = 2$ with $\beta 2 = 0$.

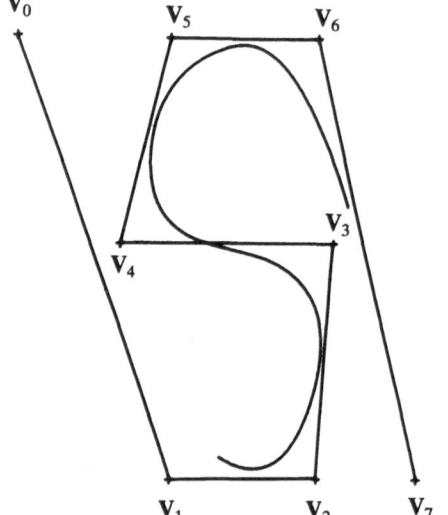

**Fig. 17.3.**   $\beta 1 = 0.5$ with $\beta 2 = 0$.

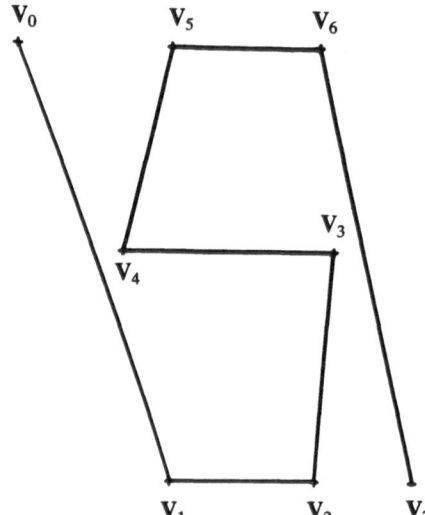

**Fig. 17.4.**   $\beta 1 = \beta 2 = 0$.

## 17.3 $\beta 2$: Tension

Substituting equation (7.11) into equation (7.2) and rearranging yields the following expression for the joint at $\mathbf{Q}_{i+1}(0) = \mathbf{Q}_i(1)$:

$$\frac{1}{\delta}[2\ \beta 1^3 \mathbf{V}_{i-1} + (4\ \beta 1^2 + 4\ \beta 1 + \beta 2)\mathbf{V}_i + 2\mathbf{V}_{i+1}]\ . \tag{17.2}$$

Consider holding $\beta 1$ and the control vertices constant. Expression (17.2) for this joint can be rewritten as the following function of $\beta 2$:

$$\frac{1}{k + \beta 2}[\mathbf{K}_i + \beta 2\mathbf{V}_i] \tag{17.3}$$

where

$$\mathbf{K}_i = 2\ \beta 1^3 \mathbf{V}_{i-1} + 4\ \beta 1(\beta 1 + 1)\mathbf{V}_i + 2\mathbf{V}_{i+1}$$

and

$$k = \delta - \beta 2 = 2\ \beta 1^3 + 4\ \beta 1^2 + 4\ \beta 1 + 2\ .$$

Now construct the vector from vertex $\mathbf{V}_i$ to this joint:

$$\frac{1}{k + \beta 2}[\mathbf{K}_i + \beta 2\mathbf{V}_i] - \mathbf{V}_i = \frac{1}{k + \beta 2}[\mathbf{K}_i - k\mathbf{V}_i]\ . \tag{17.4}$$

Several observations can be made from this form (17.4). Large values of $\beta 2$ will yield a small distance between the vertex $\mathbf{V}_i$ and this joint; thus, increasing $\beta 2$ has the effect of *attracting* the joint to the corresponding control vertex. This same effect can also be produced by small negative values of $\beta 2$ (large in absolute value). On the other hand, as the value of $\beta 2$ approaches $-k$, the distance between the vertex and the joint becomes indefinitely large; that is, the joint is *repelled* from the corresponding control vertex. Observe also that the vector from the control vertex to the joint given by expression (17.4) is independent of $\beta 2$ except for one term in the denominator. This means that this vector has a constant direction independent of $\beta 2$; thus, modifying $\beta 2$ changes only the distance between the control vertex and the joint, not the direction of the vector between them. Hence, adjusting the value of $\beta 2$ at a joint has the effect of attracting or repelling this joint towards the corresponding control vertex, and the movement of the joint is along a straight line.

The $\beta 2$ shape parameter provides the mechanism for the mathematical modeling of tension in the Beta-spline. Intuitively, $\beta 2$ measures the "tightness" or "looseness" of the curve. Large values of $\beta 2$ engender tight curves with fewer inflection points. Figures 17.5 through 17.10 show Beta-spline curves with uniform shape parameter values of $\beta 1 = 1$ with $\beta 2 = 5$, $\beta 2 = 10$, $\beta 2 = 20$, $\beta 2 = 50$, $\beta 2 = 100$, and $\beta 2 = 200$, respectively.

Negative values of $\beta 2$ produce interesting effects. As explained above, the effects are different for $-k < \beta 2 < 0$ and $\beta 2 < -k$ (for $\beta 1 = 1$, $k = 12$). As $\beta 2$ varies from 0 to $-k$, each joint is repelled from its corresponding control vertex, and the curve becomes increasingly loose with the introduction of additional inflection points.

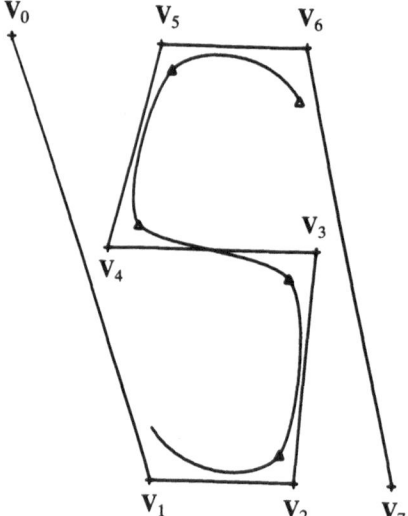

**Fig. 17.5.**  $\beta 1 = 1$ with $\beta 2 = 5$.

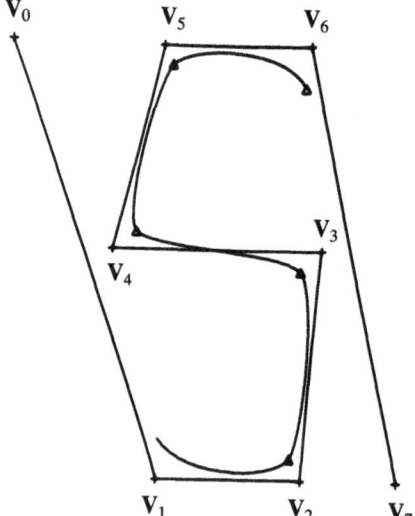

**Fig. 17.6.**  $\beta 1 = 1$ with $\beta 2 = 10$.

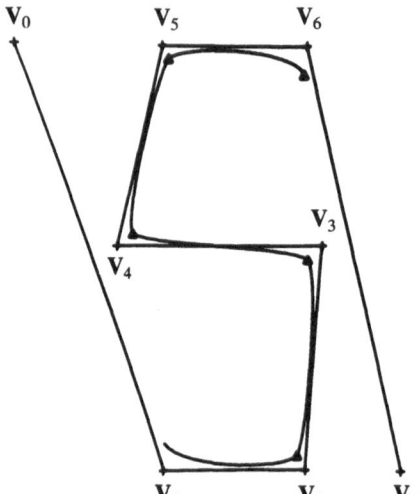

**Fig. 17.7.**  $\beta 1 = 1$ with $\beta 2 = 20$.

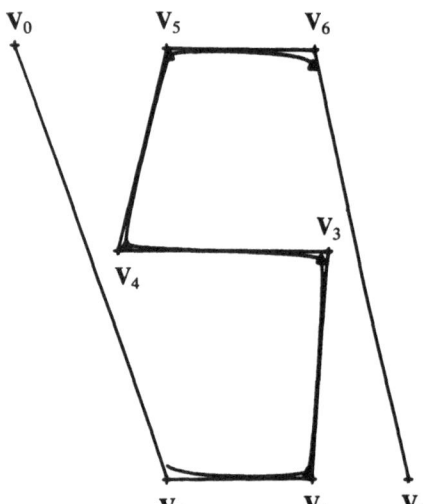

**Fig. 17.8.**  $\beta 1 = 1$ with $\beta 2 = 50$.

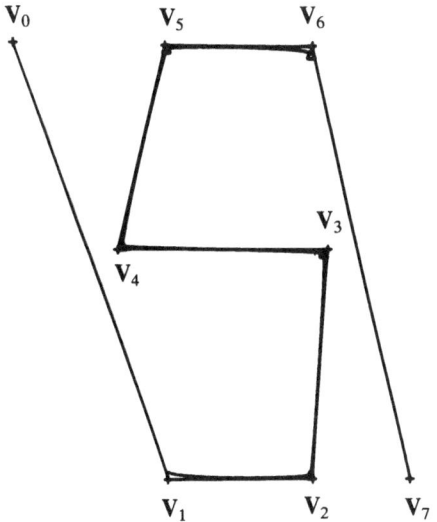

**Fig. 17.9.**  $\beta1 = 1$ with $\beta2 = 100$.

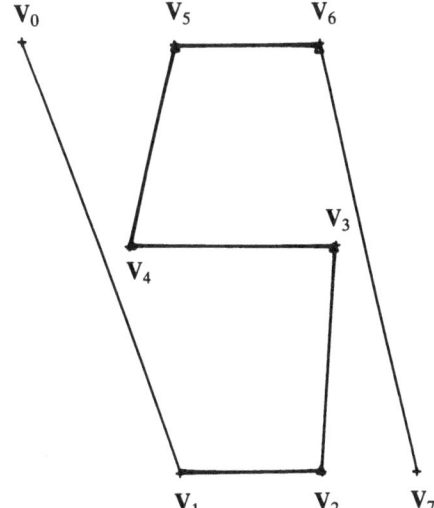

**Fig. 17.10.**  $\beta1 = 1$ with $\beta2 = 200$.

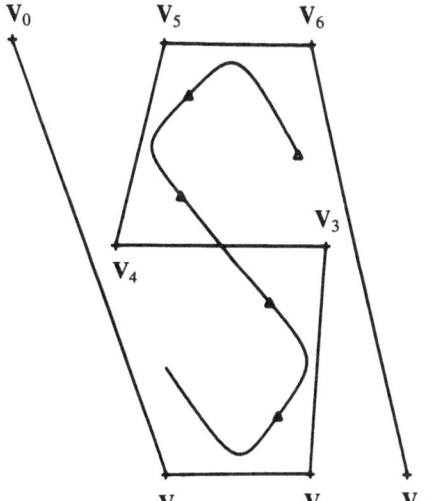

**Fig. 17.11.**  $\beta1 = 1$ with $\beta2 = -4$.

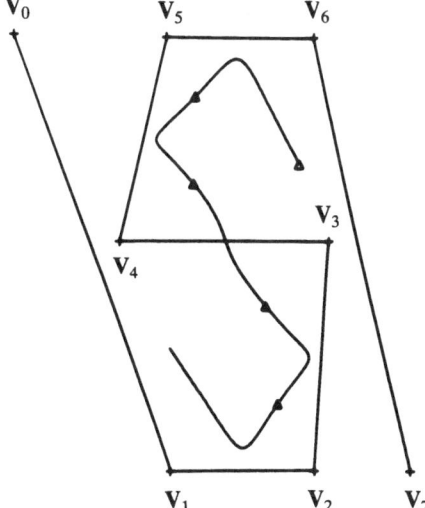

**Fig. 17.12.**  $\beta1 = 1$ with $\beta2 = -5$.

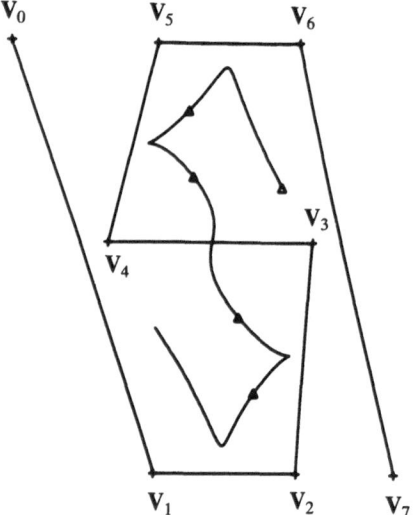

**Fig. 17.13.**   $\beta 1 = 1$ with $\beta 2 = -6$.

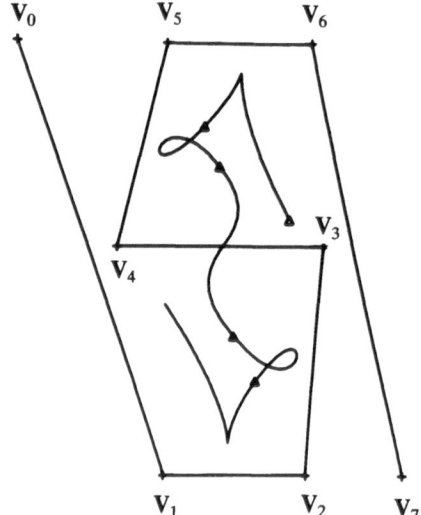

**Fig. 17.14.**   $\beta 1 = 1$ with $\beta 2 = -7$.

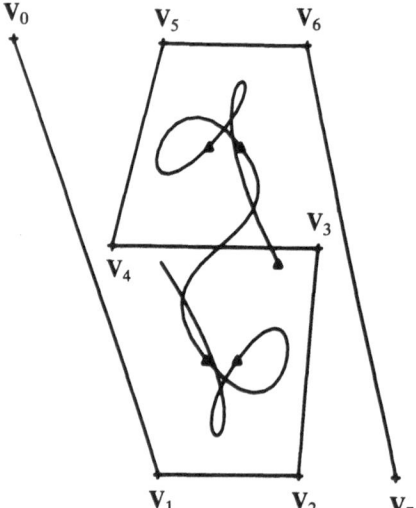

**Fig. 17.15.**   $\beta 1 = 1$ with $\beta 2 = -8$.

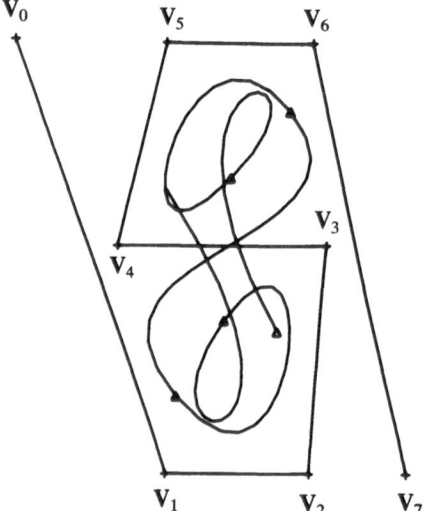

**Fig. 17.16.**   $\beta 1 = 1$ with $\beta 2 = -9$.

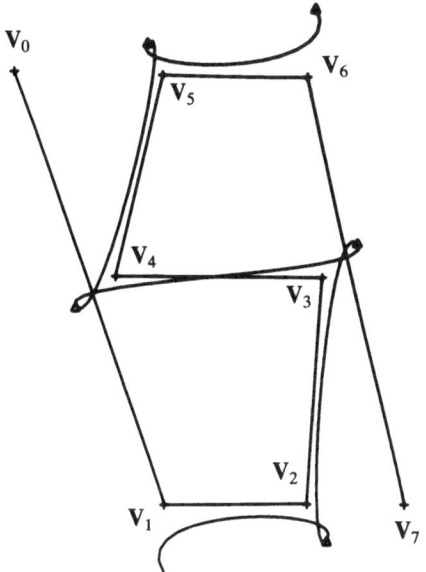

**Fig. 17.17.**   $\beta 1 = 1$ with $\beta 2 = -25$.

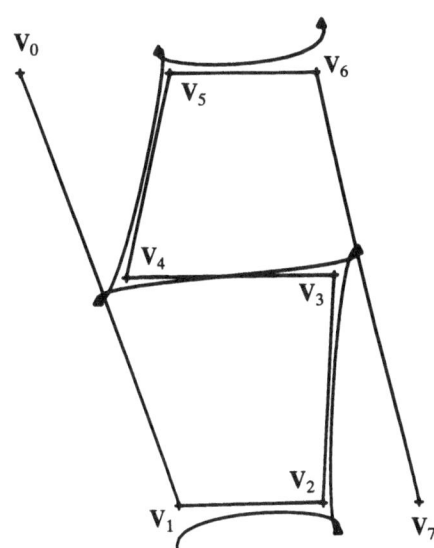

**Fig. 17.18.**   $\beta 1 = 1$ with $\beta 2 = -30$.

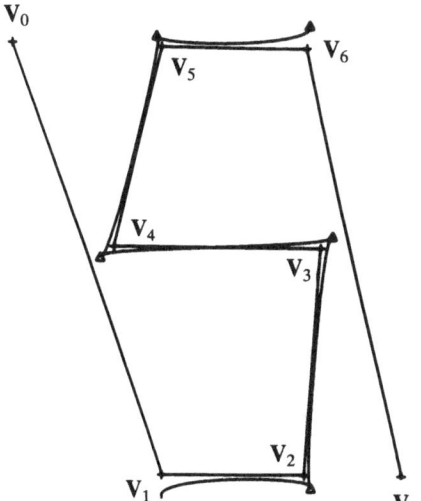

**Fig. 17.19.**   $\beta 1 = 1$ with $\beta 2 = -50$.

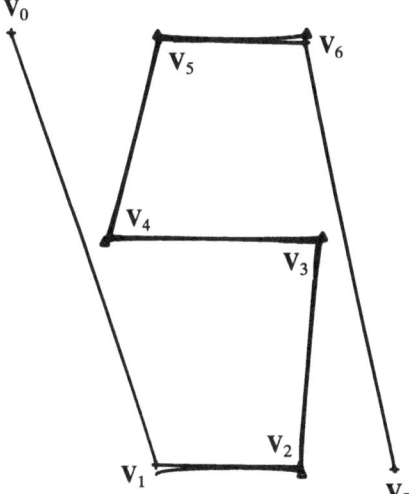

**Fig. 17.20.**   $\beta 1 = 1$ with $\beta 2 = -100$.

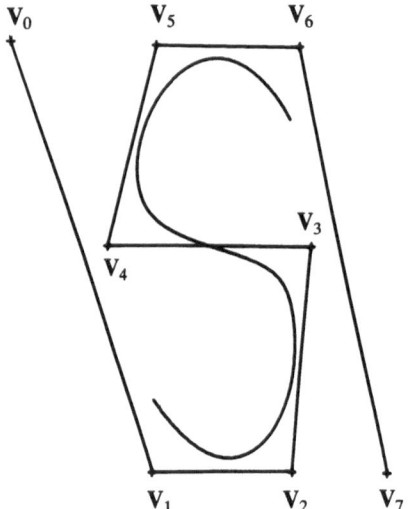

**Fig. 17.21.** $\beta 1 = 1$ with $\beta 2 = 2$ at fourth vertex.

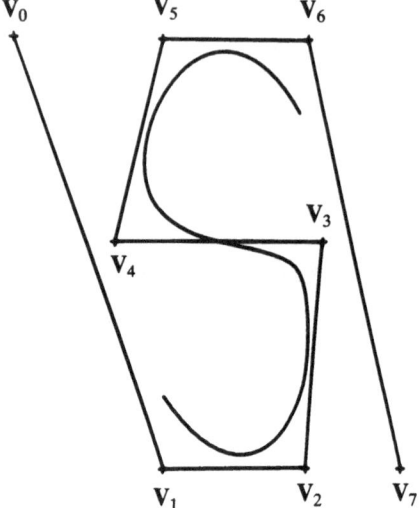

**Fig. 17.22.** $\beta 1 = 1$ with $\beta 2 = 5$ at fourth vertex.

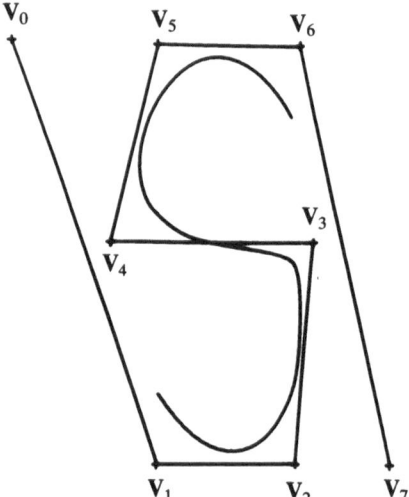

**Fig. 17.23.** $\beta 1 = 1$ with $\beta 2 = 10$ at fourth vertex.

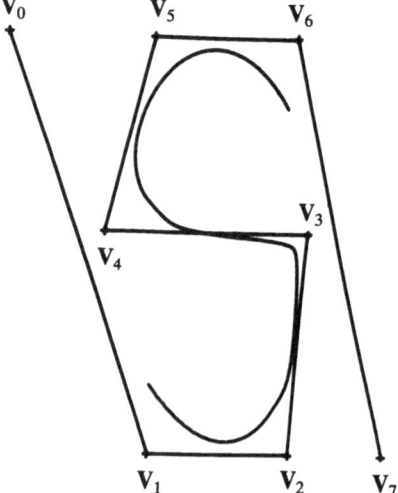

**Fig. 17.24.** $\beta 1 = 1$ with $\beta 2 = 20$ at fourth vertex.

This is shown in Figures 17.11 through 17.16 with $\beta2 = -4$, $\beta2 = -5$, $\beta2 = -6$, $\beta2 = -7$, $\beta2 = -8$, and $\beta2 = -9$, respectively.

As $\beta2$ continues to become increasingly negative, each joint is attracted to its corresponding control vertex, but from the opposite side of the control polygon. Figures 17.17 through 17.20 show $\beta2 = -25$, $\beta2 = -30$, $\beta2 = -50$, and $\beta2 = -100$, respectively.

## 17.4  β2: Continuous Shape Parameters

The shape parameters permit shape control over and above that provided by control vertex specification. Figures 17.21 through 17.24 show the curve with uniform $\beta1 = 1$ and $\beta2 = 0$ at all the joints except the one that corresponds to the fourth vertex. At this joint, $\beta2 = 2$, $\beta2 = 5$, $\beta2 = 10$, and $\beta2 = 20$ in these figures, respectively.

It should also be mentioned that it is desirable to have $\beta1$ and $\beta2$ values which are not wildly disparate. Remember that the Beta-spline contains the uniform B-spline as a *special case*. The additional flexibility of shape parameters is realized through a *relaxation* of the conventional continuity constraints, and as such the Beta-spline has less continuity than its B-spline counterpart. Although the Beta-spline permits the discontinuity of derivatives while maintaining the continuity of the fundamental geometric measures of unit tangent and curvature vectors, it in no way prevents the occurrence of local degeneracies (like cusps).

# 18 Controlling Surfaces Using the Shape Parameters

Analogous to the demonstration of Beta-spline curves in Chapter 17, this chapter illustrates the use of the shape parameters in controlling surfaces. The effect of bias is shown in Figures 18.1, 18.2, and 18.3. These figures show Beta-spline surfaces with uniform shape parameter values of $\beta 1 = 1$ with $\beta 2 = 0$, of $\beta 1 = 2$ with $\beta 2 = 0$, and of $\beta 1 = 0.5$ with $\beta 2 = 0$, respectively. A Beta-spline surface where both $\beta 1$ and $\beta 2$ are zero is illustrated in Figure 18.4. Increasing uniform tension values of $\beta 2 = 5$, $\beta 2 = 10$, $\beta 2 = 20$, $\beta 2 = 50$, $\beta 2 = 100$, and $\beta 2 = 200$ with a uniform bias value of $\beta 1 = 1$ are shown in Figures 18.5 through 18.10. Negative tension values greater than $-k$ are illustrated in Figures 18.11 through 18.16 where $\beta 2 = -4$, $\beta 2 = -5$, $\beta 2 = -6$, $\beta 2 = -7$, $\beta 2 = -8$, and $\beta 2 = -9$, respectively, with $\beta 1 = 1$. Tension values which are more negative (that is, less than $-k$) are shown in Figures 18.17 through 18.20, where $\beta 2 = -25$, $\beta 2 = -30$, $\beta 2 = -50$, and $\beta 2 = -100$, respectively, with $\beta 1 = 1$. Continuous shape parameters are illustrated in Figures 18.21 through 18.24. Here, $\beta 1$ is uniformly unity and $\beta 2$ is zero at all the joints except the one that corresponds to the top vertex. At the joint, $\beta 2 = 2$, $\beta 2 = 5$, $\beta 2 = 10$, and $\beta 2 = 20$ in Figures 18.21 through 18.24, respectively.

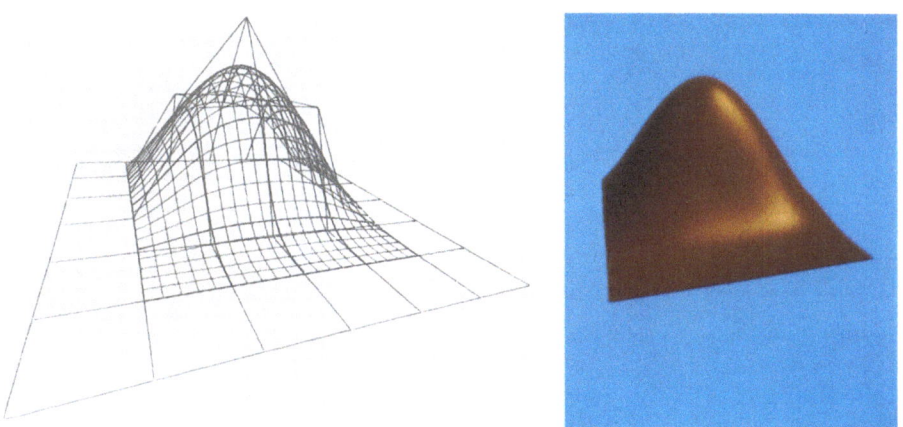

**Fig. 18.1.** $\beta 1 = 1$ with $\beta 2 = 0$.

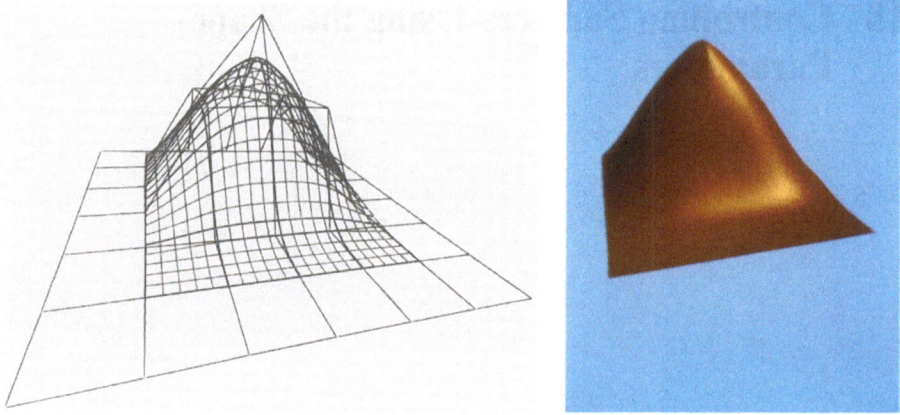

**Fig. 18.2.**   $\beta 1 = 2$ with $\beta 2 = 0$.

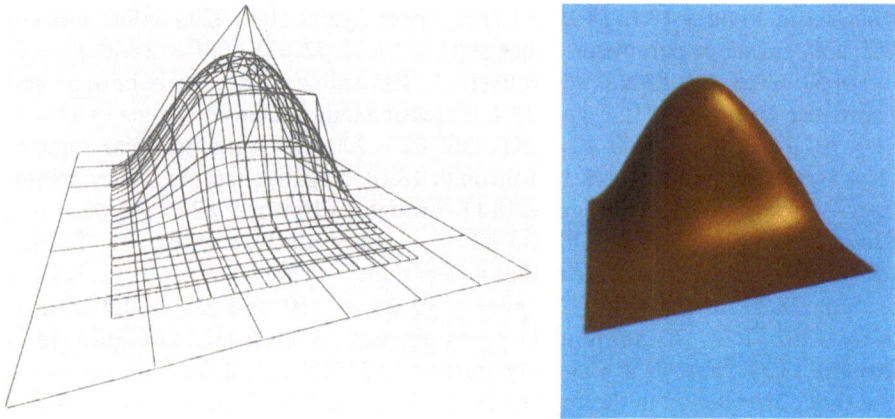

**Fig. 18.3.**   $\beta 1 = 0.5$ with $\beta 2 = 0$.

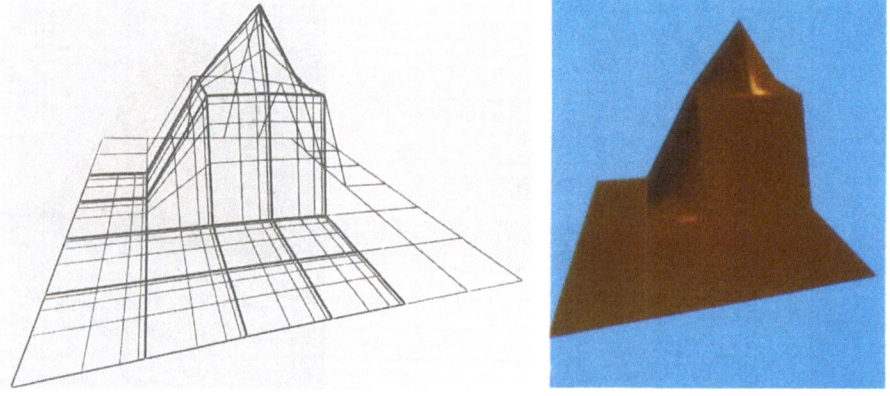

**Fig. 18.4.**   $\beta 1 = 0.0001$ with $\beta 2 = 0$.

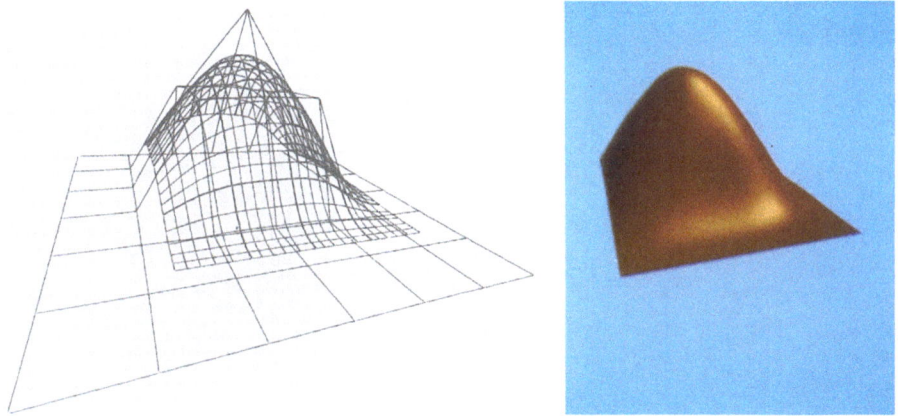

**Fig. 18.5.** $\beta 1 = 1$ with $\beta 2 = 5$.

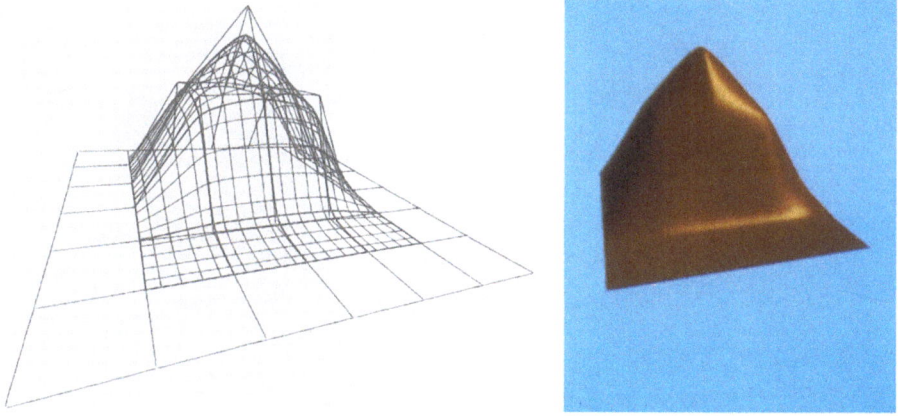

**Fig. 18.6.** $\beta 1 = 1$ with $\beta 2 = 10$.

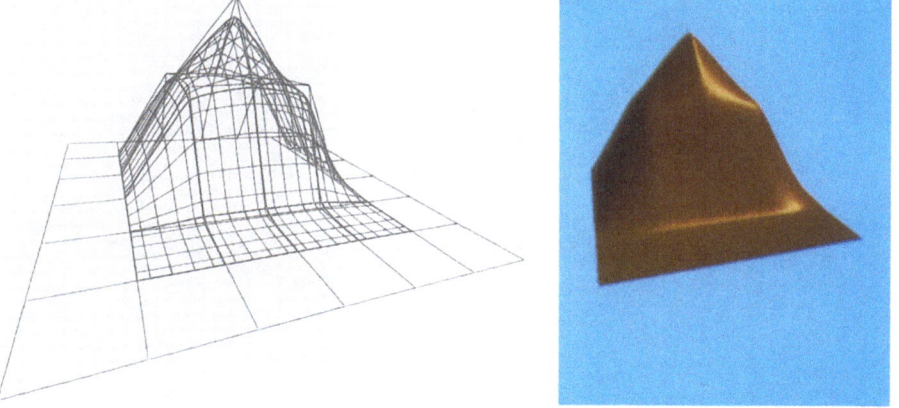

**Fig. 18.7.** $\beta 1 = 1$ with $\beta 2 = 20$.

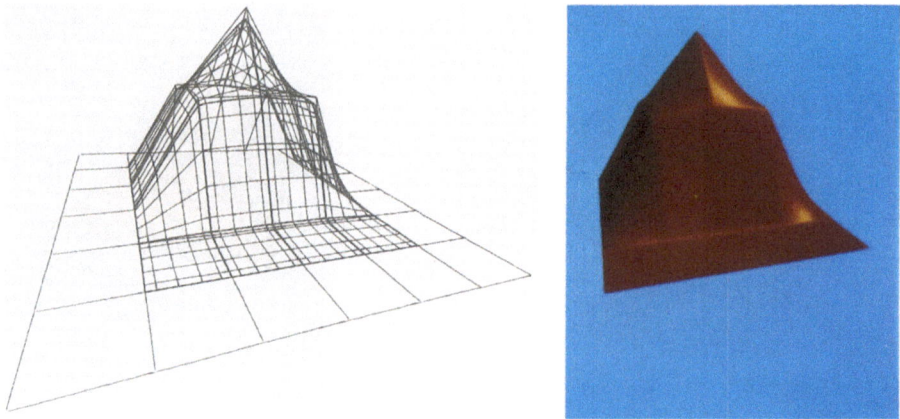

**Fig. 18.8.**  $\beta 1 = 1$ with $\beta 2 = 50$.

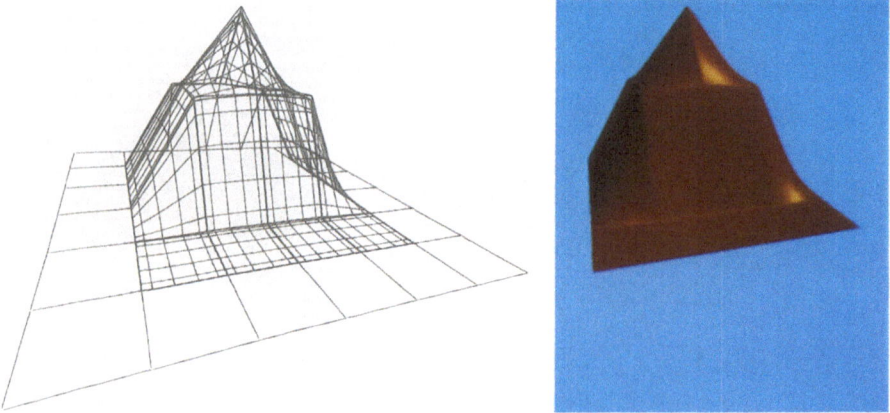

**Fig. 18.9.**  $\beta 1 = 1$ with $\beta 2 = 100$.

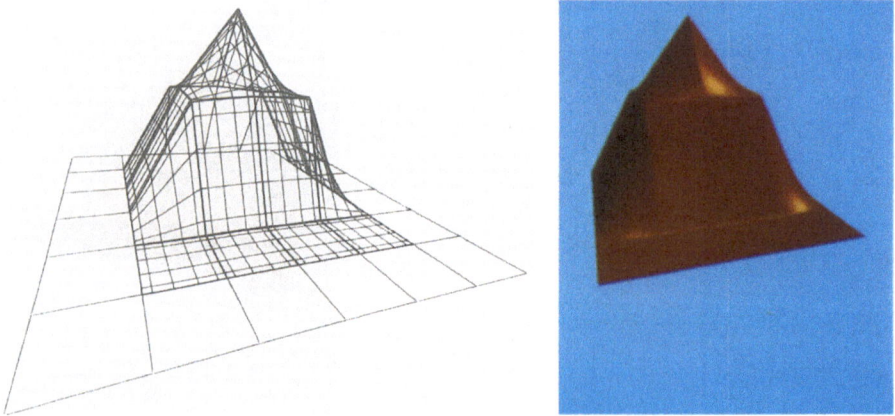

**Fig. 18.10.**  $\beta 1 = 1$ with $\beta 2 = 200$.

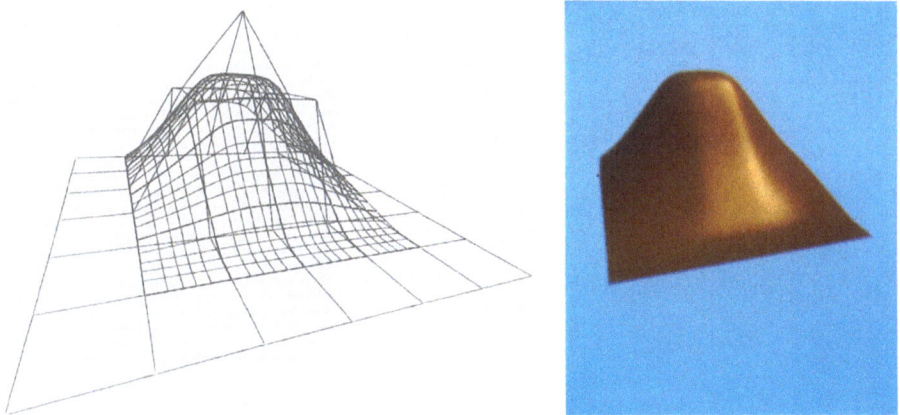

**Fig. 18.11.**   $\beta 1 = 1$ with $\beta 2 = -4$.

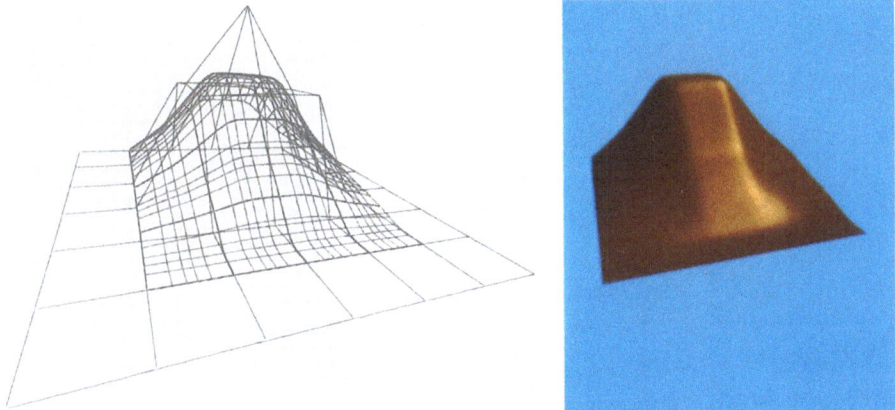

**Fig. 18.12.**   $\beta 1 = 1$ with $\beta 2 = -5$.

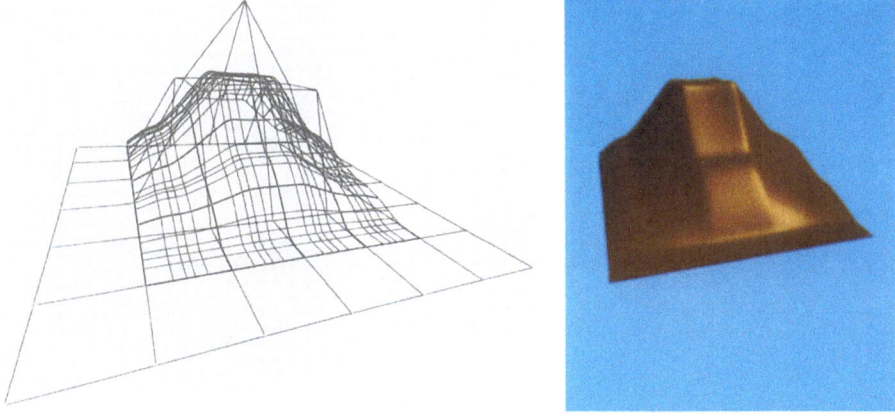

**Fig. 18.13.**   $\beta 1 = 1$ with $\beta 2 = -6$.

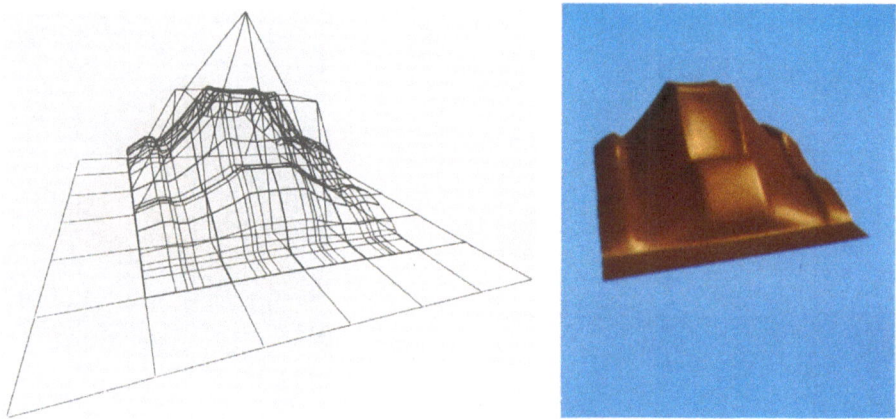

**Fig. 18.14.**   $\beta1 = 1$ with $\beta2 = -7$.

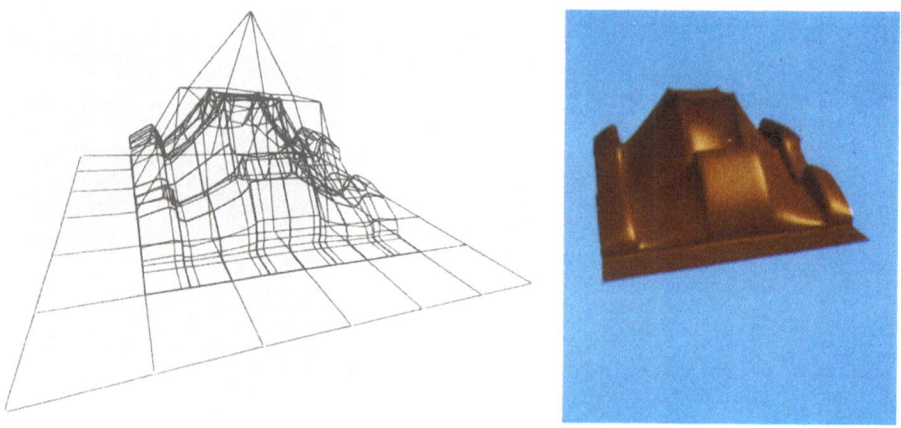

**Fig. 18.15.**   $\beta1 = 1$ with $\beta2 = -8$.

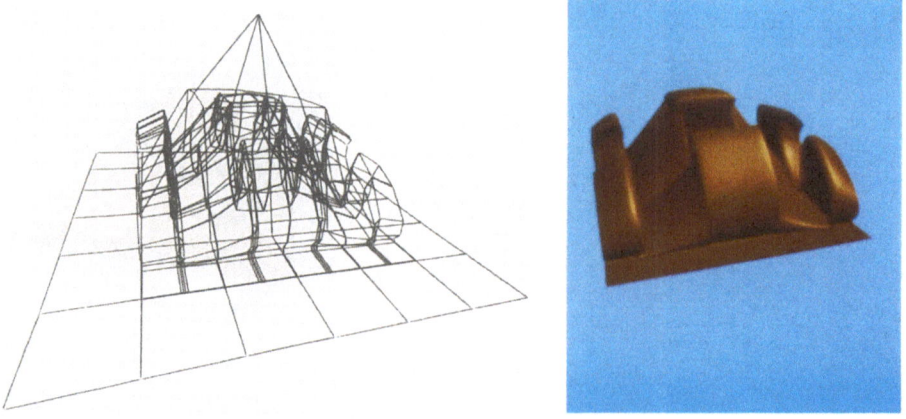

**Fig. 18.16.**   $\beta1 = 1$ with $\beta2 = -9$.

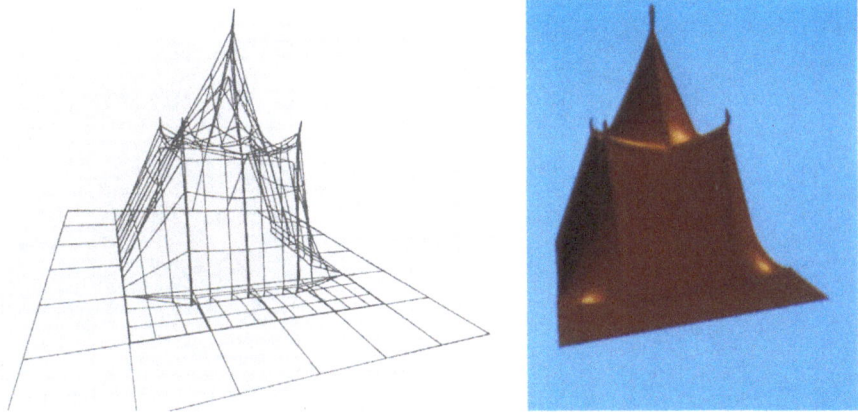

**Fig. 18.17.** $\beta 1 = 1$ with $\beta 2 = -25$.

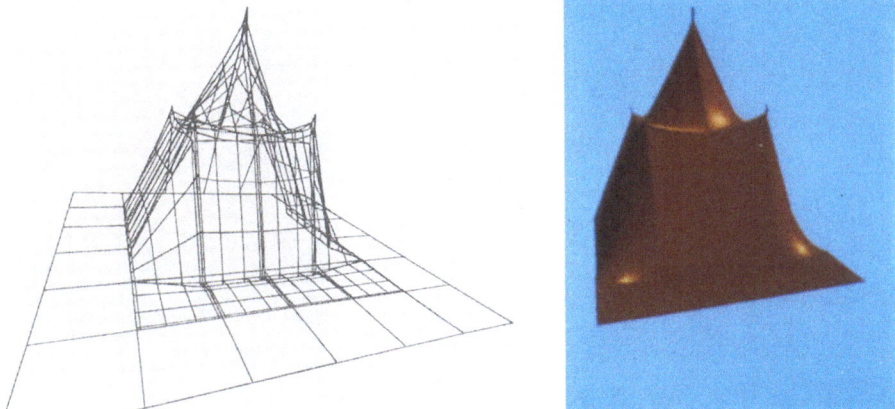

**Fig. 18.18.** $\beta 1 = 1$ with $\beta 2 = -30$.

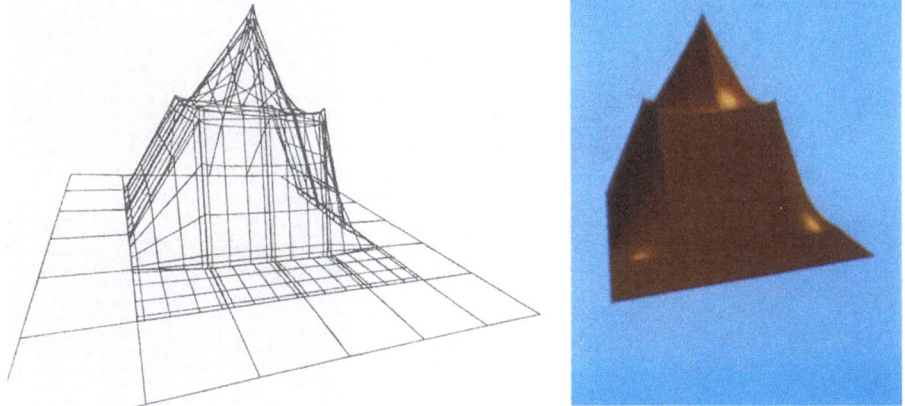

**Fig. 18.19.** $\beta 1 = 1$ with $\beta 2 = -50$.

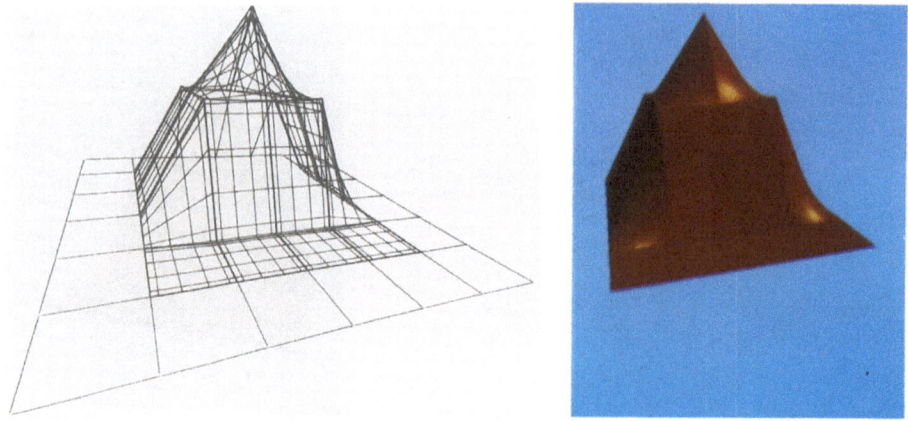

**Fig. 18.20.**   $\beta1 = 1$ with $\beta2 = -100$.

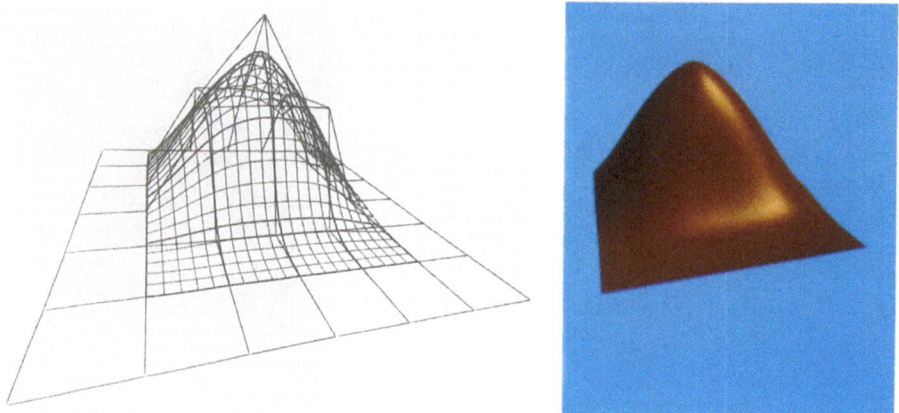

**Fig. 18.21.**   $\beta1 = 1$ with $\beta2 = 2$ at top vertex.

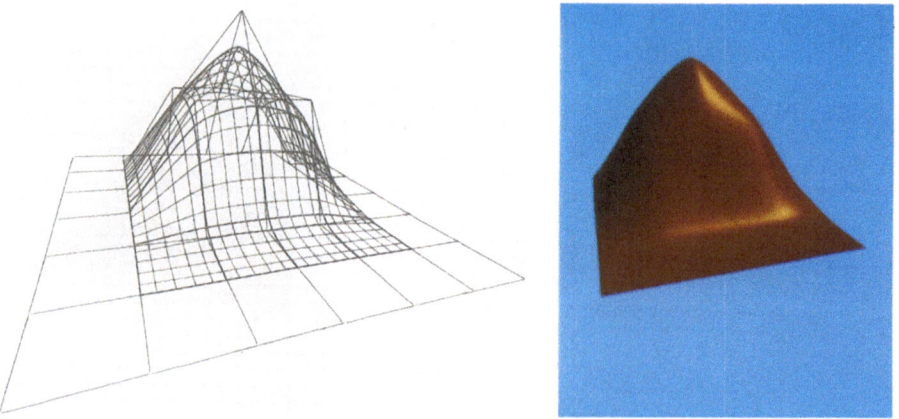

**Fig. 18.22.**   $\beta1 = 1$ with $\beta2 = 5$ at top vertex.

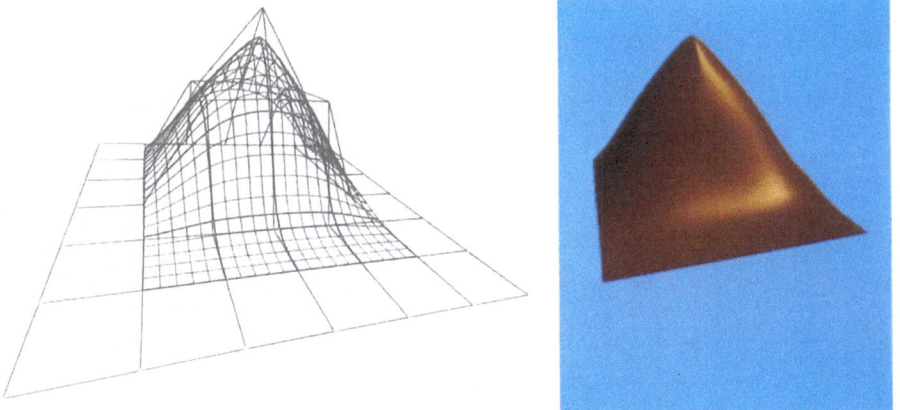

**Fig. 18.23.** $\beta 1 = 1$ with $\beta 2 = 10$ at top vertex.

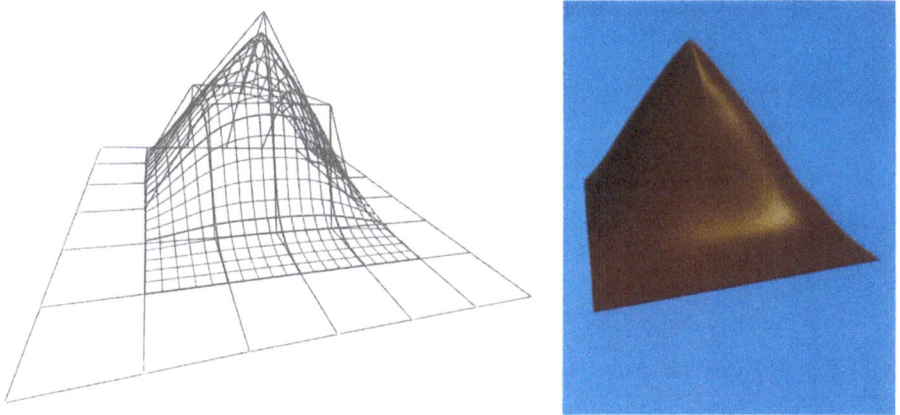

**Fig. 18.24.** $\beta 1 = 1$ with $\beta 2 = 20$ at top vertex.

# 19 Synthetic Images of Beta-spline Objects

In computer graphics, it is of interest to *render* a Beta-spline, that is, create a realistic two-dimensional image representing three-dimensional Beta-spline objects. This chapter shows several synthetic color images of Beta-spline objects including such features as specular highlights and texture patterns.

Figure 19.1 is an array based on a single Beta-spline bottle with only the tension varying. Each sub-image is defined by the same set of control vertices. All that changes between sub-images is the value of tension. It is increasing from left to right

**Fig. 19.1.** Beta-spline bottle with increasing tension values and different textures. (This image first appeared on the back cover of the *ACM/SIGGRAPH '83 Conference Proceedings.*)

and from top to bottom in an exponential fashion. Each of the tensed bottles has different material characteristics including metallic, dusty and plastic substances. The bias remains fixed throughout. Figure 19.2 shows the same scene without the mapping of texture patterns onto the objects.

Figure 19.3 shows three Beta-spline ornaments. Two of the objects occur multiply with three different values of tension. The ornaments are constructed from various glossy materials. The bias remains fixed throughout.

Figure 19.4 is an array that is based on a single, shiny, textured Beta-spline bottle. Both the tension and bias are varied. The tension increases from left to right and the bias increases from top to bottom. Figure 19.5 shows an array of three different Beta-spline objects with tension increasing from left to right and bias fixed. The material characteristics of the objects also vary including metallic and plastic substances. The objects increase in complexity from top to bottom with the bottom-most object exhibiting subtle shape changes due to tension.

Figure 19.6 shows five Beta-spline pewter goblets all defined by the same control vertices. The changing shapes from round to flat are controlled by increasing tension values. The values of tension from left to right are: 0, 5, 10, 20, and 50.

The scene in Figure 19.7 is composed entirely of Beta-spline surfaces. The objects have varying material properties, including metallic, dusty, and plastic substances. Prefiltered and stochastically generated textures are also applied to some of the objects. The bias for the surfaces remains fixed throughout with the tension varying.

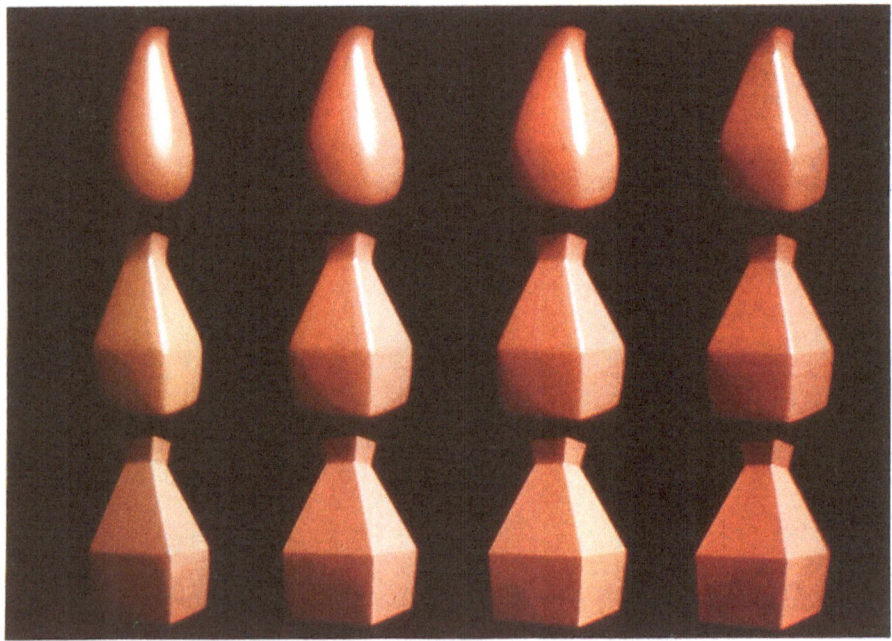

**Fig. 19.2.**   Beta-spline bottle with increasing tension.

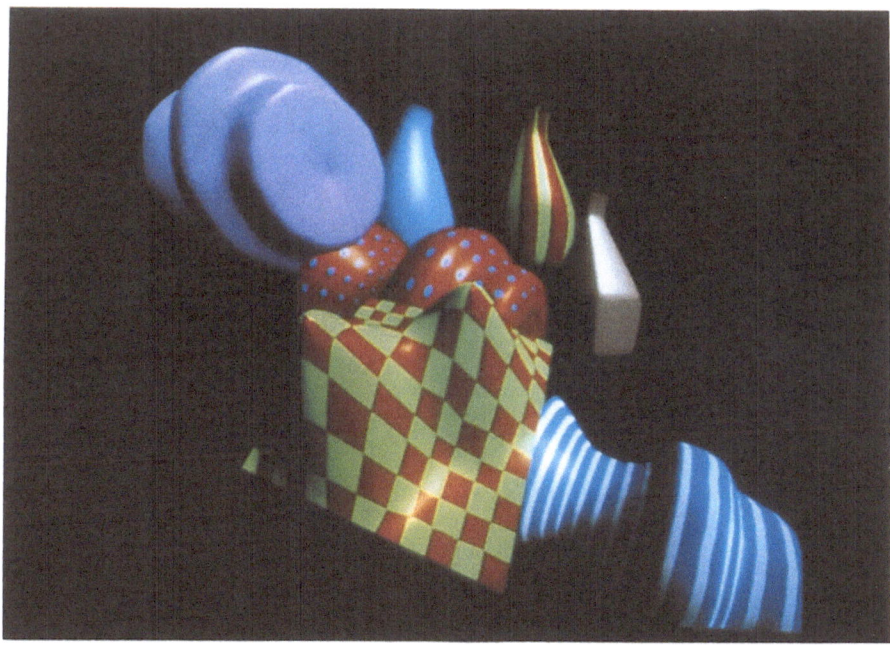

**Fig. 19.3.** "Christmas at Macy's". (This image first appeared on the front cover of the 27 June 1983 issue of *Nikkei Computer*.)

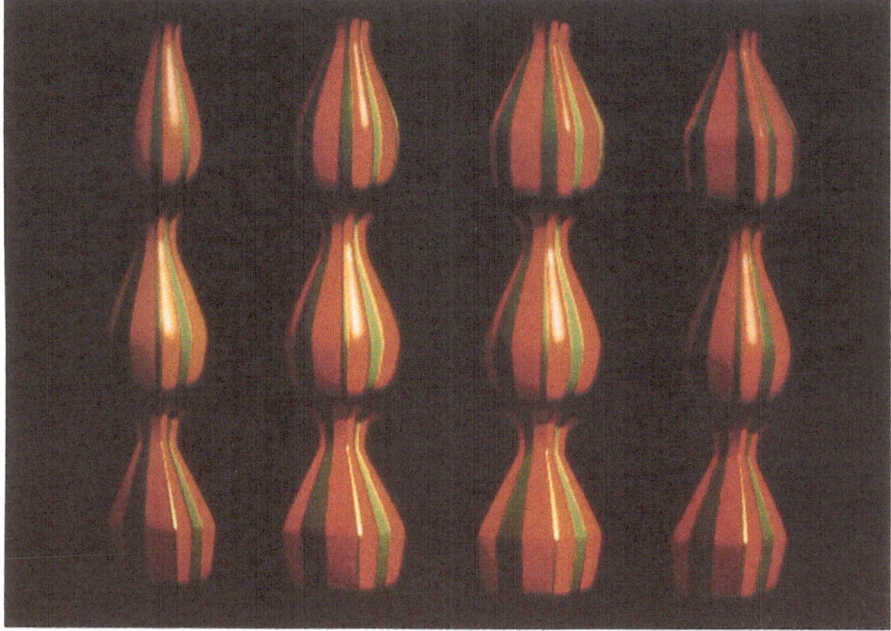

**Fig. 19.4.** Beta-spline bottle with different bias and tension values. (This image first appeared on page 119 in *Computer Graphics* edited by Joan Scott, Gulf Publishing Company, 1984.)

**Fig. 19.5.** Three Beta-spline objects with increasing tension values and different textures. (This image first appeared on the front cover of the 9-10/1983 issue of *PIXEL*, No. 14.)

**Fig. 19.6.** Pewter goblets with different tension values. (This image first appeared on the front cover of the 5/1984 issue of *PIXEL*, No. 20.)

**Fig. 19.7.**   Scene composed entirely of Beta-spline surfaces.

# 20 Conclusion and Future Research Directions

The Beta-spline is a mathematical technique for curve and surface representation that has been developed expressly for geometrical and graphical applications. The formulation is based on the constraints of continuous fundamental geometric measures. Elementary differential geometry concepts were reviewed and the *unit tangent* and *curvature* vectors replaced parametric derivative vectors in the continuity constraints. In addition to being more appropriate, the use of these geometric measures adds degrees of freedom which can be captured to provide further control of shape, via inherent *shape parameters*, $\beta 1$ and $\beta 2$. These shape parameters have the property that $\beta 1 = 1$ indicates continuity of the parametric first derivative vector and $\beta 1 = 1$ with $\beta 2 = 0$ indicates continuity of the parametric first and second derivative vectors. Since these constraints are expressed in terms of the shape parameters, the derivation of the Beta-spline curve representation utilized symbolic, not numeric, computation. Explicit expressions were provided for the Beta-spline as well as for various derivatives.

Two methods were designed and analyzed for the evaluation and perturbation of a Beta-spline with uniform (assuming a single value over the entire curve) shape parameters. This assumption was then generalized so that the shape parameters are continuous, each varying continuously along the curve, and methods were designed and analyzed for the evaluation and perturbation of the Beta-spline with continuous shape parameters. End conditions for Beta-spline curves were classified and analyzed.

The tensor product Beta-spline surface was then explained. Methods were designed and analyzed for its evaluation and perturbation with uniform shape parameters. The surface representation was generalized for continuous shape parameters, and methods for the evaluation and perturbation in this case were designed and analyzed. Boundary conditions for Beta-spline surfaces are classified and analyzed.

The computational requirements of the evaluation and perturbation algorithms for the Beta-spline were found to be competitive with those for the B-spline as given in [2]. Finally, the geometrical interpretation of these shape parameters and their relation to the mathematical modeling of tension was also investigated.

There are many directions in which the Beta-spline can be further generalized. For example, the continuous shape parameter function could be more complex so as to be dependent on more than the present two (for curves) or four (for surfaces) values of the discrete shape parameters. Another idea is the development of a family of such splines, the members of which would each have different kinds of shape parameters which would be appropriate for different applications. In addition, higher order splines which would have more than the present two kinds of shape parameters could be investigated. Also, the Beta-spline representation could be extended to include the case of non-uniformly spaced or multiple parametric knot values. Finally, there are additional end and boundary conditions which could be explored for the Beta-spline.

Since the Beta-spline is a new representation, it will engender a wide range of future research. From a mathematical point of view, the fundamental approximation theoretic implications of the Beta-spline have yet to be explored. For example, it would be of interest to derive the variational principle which corresponds to the Beta-spline. For computer graphics, rendering algorithms for the creation and display of Beta-spline objects will be developed. In the computer aided design field, its utility as a three-dimensional object representation should be investigated. In terms of hardware, it would be very desirable to design evaluation and perturbation algorithms using parallel processing techniques and to build a "Beta-spline box" based on this multiprocessor architecture. A particularly likely direction of immediate research would be the development of subdivision techniques for Beta-splines. The technique of subdivision has been employed in computer aided geometric design and modeling to provide good design handles, and in computer graphics to generate smooth models for display in a kind of divide and conquer approach [21].

The combination of the geometric nature of this technique with shape control via the shape parameters as well as the presence of local control capability forms a powerful representation which will be useful in computer graphics and computer aided geometric design and modeling.

# Appendix: REDUCE Programs

## A.1 REDUCE Program to Determine the Coefficient Functions

```
OFF ECHO$
MATRIX B(13, 13), RHS(13, 1), RESULT(13, 1)$

FACTOR1 := BETA1 * BETA1 + BETA2;
FACTOR2 := 2 * BETA1 * BETA1 + BETA2;

B(1, 1) := 1$
B(1, 2) := 1$
B(1, 3) := 1$
B(1, 4) := 1$
B(2, 1) := − 1$
B(2, 5) := 1$
B(2, 6) := 1$
B(2, 7) := 1$
B(2, 8) := 1$
B(3, 5) := − 1$
B(3, 9) := 1$
B(3, 10) := 1$
B(3, 11) := 1$
B(3, 12) := 1$
B(4, 9) := − 1$
B(4, 13) := 1$

B(5, 2) := BETA1$
B(5, 3) := 2 * BETA1$
B(5, 4) := 3 * BETA1$

B(6, 2) := − 1$
B(6, 6) := BETA1$
B(6, 7) := 2 * BETA1$
B(6, 8) := 3 * BETA1$
```

```
B(7, 6) := −1$
B(7, 10) := BETA1$
B(7, 11) := 2 * BETA1$
B(7, 12) := 3 * BETA1$


B(8, 10) := −1$
B(8, 13) := 3 * BETA1$


B(9, 2) := BETA2$
B(9, 3) := 2 * FACTOR1$
B(9, 4) := 3 * FACTOR2$


B(10, 3) := −2$
B(10, 6) := BETA2$
B(10, 7) := 2 * FACTOR1$
B(10, 8) := 3 * FACTOR2$


B(11, 7) := −2$
B(11, 10) := BETA2$
B(11, 11) := 2 * FACTOR1$
B(11, 12) := 3 * FACTOR2$


B(12, 11) := −2$
B(12, 13) := 3 * FACTOR2$


B(13, 1) := 1$
B(13, 5) := 1$
B(13, 9) := 1$


RHS(13, 1) := 1$


ON EXP;
ON GCD;
OUT B.ANS$
OFF NAT$


WRITE "MATRIX RESULT(13, 1)$"$
RESULT := 1/B * RHS$
DENOM := DEN RESULT(1, 1);
FOR I := 1 : 13 DO RESULT(I, 1) := DENOM * RESULT(I, 1)$
RESULT := RESULT;


SHUT B.ANS$


FOR I := 1 : 13 DO STRUCTR RESULT(I, 1)$
```

BETA1 := 1;
BETA2 := 0;

RESULT := RESULT;
END$


## A.2 REDUCE Output of Coefficient Functions

MATRIX RESULT(13, 1)$$

DENOM := 2 * BETA1 ** 3 + 4 * BETA1 ** 2 + 4 * BETA1 + BETA2 + 2$

RESULT(1, 1) := 2 * BETA1 ** 3$
RESULT(2, 1) := −6 * BETA1 ** 3$
RESULT(3, 1) := 6 * BETA1 ** 3$
RESULT(4, 1) := −2 * BETA1 ** 3$
RESULT(5, 1) := 4 * BETA1 ** 2 + 4 * BETA1 + BETA2$
RESULT(6, 1) := 6 * BETA1 * (BETA1 ** 2 − 1)$
RESULT(7, 1) := 3 * (−2 * BETA1 ** 3 − 2 * BETA1 ** 2 − BETA2)$
RESULT(8, 1) := 2 * (BETA1 ** 3 + BETA1 ** 2 + BETA1 + BETA2)$
RESULT(9, 1) := 2$
RESULT(10, 1) := 6 * BETA1$
RESULT(11, 1) := 3 * (2 * BETA1 ** 2 + BETA2)$
RESULT(12, 1) := −2 * (BETA1 ** 2 + BETA1 + BETA2 + 1)$
RESULT(13, 1) := 2$


## A.3 REDUCE Program to Verify the Geometric Continuity Constraints

OFF ECHO;
MATRIX b0(4, 1), b1(4, 1), d0(4, 1), d1(4, 1), e0(4, 1), e1(4, 1);

ON EXP;
ON GCD;
FACTOR V1, V2, V3, V4, V5;
OUT CHECK.ANS;
OFF NAT;

denom := 2 * beta1 ** 3 + 4 * beta1 ** 2 + 4 * beta1 + beta2 + 2;

b0(1, 1) := 2 * beta1 ** 3/denom;
b0(2, 1) := (4 * beta1 ** 2 + 4 * beta1 + beta2)/denom;
b0(3, 1) := 2/denom;
b0(4, 1) := 0;

```
b1 (1, 1) := 0;
b1 (2, 1) := 2 * beta1 ** 3/denom;
b1 (3, 1) := (4 * beta1 ** 2 + 4 * beta1 + beta2)/denom;
b1 (4, 1) := 2/denom;


d0 (1, 1) := −6 * beta1 ** 3/denom;
d0 (2, 1) := 6 * beta1 * (beta1 ** 2 − 1)/denom;
d0 (3, 1) := 6 * beta1/denom;
d0 (4, 1) := 0;


d1 (1, 1) := 0;
d1 (2, 1) := −6 * beta1 ** 2/denom;
d1 (3, 1) := 6 * (beta1 ** 2 − 1)/denom;
d1 (4, 1) := 6/denom;


e0 (1, 1) := 12 * beta1 ** 3/denom;
e0 (2, 1) := 6 * (−2 * beta1 ** 3 − 2 * beta1 ** 2 − beta2)/denom;
e0 (3, 1) := 6 * (2 * beta1 ** 2 + beta2)/denom;
e0 (4, 1) := 0;


e1 (1, 1) := 0;
e1 (2, 1) := 6 * (2 * beta1 + beta2)/denom;
e1 (3, 1) := 6 * (−2 * beta1 − beta2 − 2)/denom;
e1 (4, 1) := 12/denom;


left1 := b0 (1, 1) * V2 + b0 (2, 1) * V3 + b0 (3, 1) * V4 + b0 (4, 1) * V5;
right1 := b1 (1, 1) * V1 + b1 (2, 1) * V2 + b1 (3, 1) * V3 + b1 (4, 1) * V4;


differ := left1 − right1;


left2 := d0 (1, 1) * V2 + d0 (2, 1) * V3 + d0 (3, 1) * V4 + d0 (4, 1) * V5;
right2 := beta1 * (d1 (1, 1) * V1 + d1 (2, 1) * V2 + d1 (3, 1) * V3 + d1 (4, 1) * V4);


differ := left2 − right2;


left3 := e0 (1, 1) * V2 + e0 (2, 1) * V3 + e0 (3, 1) * V4 + e0 (4, 1) * V5;
right3 := (beta1 ** 2) * (e1 (1, 1) * V1 + e1 (2, 1) * V2 + e1 (3, 1) * V3 + e1 (4, 1) *
    V4) + beta2 * (d1 (1, 1) * V1 + d1 (2, 1) * V2 + d1 (3, 1) * V3 + d1 (4, 1) * V4);
differ := left3 − right3;


SHUT CHECK.ANS;


END;
```

## A.4 REDUCE Output Verifying Geometric Continuity Constraints

DENOM := 2 * BETA1 * * 3 + 4 * BETA1 * * 2 + 4 * BETA1 + BETA2 + 2$

B0 (1, 1) := (2 * BETA1 * * 3)/(2 * BETA1 * * 3 + 4 * BETA1 * * 2 + 4 * BETA1 +
. BETA2 + 2)$

B0 (2, 1) := (4 * BETA1 * * 2 + 4 * BETA1 + BETA2)/(2 * BETA1 * * 3 + 4 *
BETA1 * * 2 + 4 * BETA1 + BETA2 + 2)$

B0 (3, 1) := 2/(2 * BETA1 * * 3 + 4 * BETA1 * * 2 + 4 * BETA1 + BETA2 + 2)$

B0 (4, 1) := 0$

B1 (1, 1) := 0$

B1 (2, 1) := (2 * BETA1 * * 3)/(2 * BETA1 * * 3 + 4 * BETA1 * * 2 + 4 * BETA1 +
BETA2 + 2)$

B1 (3, 1) := (4 * BETA1 * * 2 + 4 * BETA1 + BETA2)/(2 * BETA1 * * 3 + 4 *
BETA1 * * 2 + 4 * BETA1 + BETA2 + 2)$

B1 (4, 1) := 2/(2 * BETA1 * * 3 + 4 * BETA1 * * 2 + 4 * BETA1 + BETA2 + 2)$

D0 (1, 1) := (−6 * BETA1 * * 3)/(2 * BETA1 * * 3 + 4 * BETA1 * * 2 + 4 *
BETA1 + BETA2 + 2)$

D0 (2, 1) := (6 * BETA1 * (BETA1 * * 2 − 1))/(2 * BETA1 * * 3 + 4 * BETA1 * *
2 + 4 * BETA1 + BETA2 + 2)$

D0 (3, 1) := (6 * BETA1)/(2 * BETA1 * * 3 + 4 * BETA1 * * 2 + 4 * BETA1 +
BETA2 + 2)$

D0 (4, 1) := 0$

D1 (1, 1) := 0$

D1 (2, 1) := (−6 * BETA1 * * 2)/(2 * BETA1 * * 3 + 4 * BETA1 * * 2 + 4 *
BETA1 + BETA2 + 2)$

D1 (3, 1) := (6 * (BETA1 * * 2 − 1))/(2 * BETA1 * * 3 + 4 * BETA1 * * 2 + 4 *
BETA1 + BETA2 + 2)$

D1 (4, 1) := 6/(2 * BETA1 * * 3 + 4 * BETA1 * * 2 + 4 * BETA1 + BETA2 + 2)$

E0 (1, 1) := (12 * BETA1 * * 3)/(2 * BETA1 * * 3 + 4 * BETA1 * * 2 + 4 * BETA1 +
BETA2 + 2)$

E0(2, 1) := (6 * (−2 * BETA1 ** 3 − 2 * BETA1 ** 2 − BETA2))/(2 * BETA1 ** 3 + 4 * BETA1 ** 2 + 4 * BETA1 + BETA2 + 2)$

E0(3, 1) := (6 * (2 * BETA1 ** 2 + BETA2))/(2 * BETA1 ** 3 + 4 * BETA1 ** 2 + 4 * BETA1 + BETA2 + 2)$

E0(4, 1) := 0$

E1(1, 1) := 0$

E1(2, 1) := (6 * (2 * BETA1 + BETA2))/(2 * BETA1 ** 3 + 4 * BETA1 ** 2 + 4 * BETA1 + BETA2 + 2)$

E1(3, 1) := (6 * (−2 * BETA1 − BETA2 − 2))/(2 * BETA1 ** 3 + 4 * BETA1 ** 2 + 4 * BETA1 + BETA2 + 2)$

E1(4, 1) := 12/(2 * BETA1 ** 3 + 4 * BETA1 ** 2 + 4 * BETA1 + BETA2 + 2)$

LEFT1 := (2 * V2 * BETA1 ** 3 + V3 * (4 * BETA1 ** 2 + 4 * BETA1 + BETA2) + 2 * V4)/(2 * BETA1 ** 3 + 4 * BETA1 ** 2 + 4 * BETA1 + BETA2 + 2)$

RIGHT := (2 * V2 * BETA1 ** 3 + V3 * (4 * BETA1 ** 2 + 4 * BETA1 + BETA2) + 2 * V4)/(2 * BETA1 ** 3 + 4 * BETA1 ** 2 + 4 * BETA1 + BETA2 + 2)$

DIFFER := 0$

LEFT2 := (−6 * V2 * BETA1 ** 3 + 6 * V3 * BETA1 * (BETA1 ** 2 − 1) + 6 * V4 * BETA1)/(2 * BETA1 ** 3 + 4 * BETA1 ** 2 + 4 * BETA1 + BETA2 + 2)$

RIGHT2 := (−6 * V2 * BETA1 ** 3 + 6 * V3 * BETA1 * (BETA1 ** 2 − 1) + 6 * V4 * BETA1)/(2 * BETA1 ** 3 + 4 * BETA1 ** 2 + 4 * BETA1 + BETA2 + 2)$

DIFFER := 0$

LEFT3 := (12 * V2 * BETA1 ** 3 + 6 * V3 * (−2 * BETA1 ** 3 − 2 * BETA1 ** 2 − BETA2) + 6 * V4 * (2 * BETA1 ** 2 + BETA2))/(2 * BETA1 ** 3 + 4 * BETA1 ** 2 + 4 * BETA1 + BETA2 + 2)$

RIGHT3 := (12 * V2 * BETA1 ** 3 + 6 * V3 * (−2 * BETA1 ** 3 − 2 * BETA1 ** 2 − BETA2) + 6 * V4 * (2 * BETA1 ** 2 + BETA2))/(2 * BETA1 ** 3 + 4 * BETA1 ** 2 + 4 * BETA1 + BETA2 + 2)$

DIFFER := 0$

# References

1. Barsky, Brian A. "Exponential and Polynomial Methods for Applying Tension to an Inter- polating Spline Curve", *Computer Vision, Graphics, and Image Processing*, Vol. 27, No. 1, July 1984, pp. 1–18.
2. Barsky, Brian A. *A Study of the Parametric Uniform B-spline Curve and Surface Representa- tions*, Tech. Report No. UCB/CSD 83/118, Computer Science Division, Electrical Engineer- ing and Computer Sciences Department, University of California, Berkeley, California, USA, May 1983.
3. Barsky, Brian A. "End Conditions and Boundary Conditions for Uniform B-spline Curve and Surface Representations", *Computers in Industry*, Vol. 3, Nos. 1 and 2, March and June 1982, pp. 17–29. Special Steven A. Coons Memorial Issue.
4. Barsky, Brian A. and Thomas, Spencer W. *TRANSPLINE Curve Representation System*, Tech. Report No. UUCS-80-104, Department of Computer Science, University of Utah, Salt Lake City, Utah, USA, April 1980.
5. Barsky, Brian A. and Thomas, Spencer W. "TRANSPLINE–A System for Representing Curves Using Transformations Among Four Spline Formulations", *The Computer Journal*, Vol. 24, No. 3, August 1981, pp. 271–277.
6. Cline, Alan K. "Scalar- and Planar-Valued Curve Fitting Using Splines Under Tension", *Communications of the ACM*, Vol. 17, No. 4, April 1974, pp. 218–220.
7. Coons, Steven A. "Surface Patches and B-spline Curves". In *Computer Aided Geometric Design*, edited by Barnhill, Robert E. and Riesenfeld, Richard F., Academic Press, New York, USA, 1974, pp. 1–16.
8. Faux, Ivor D. and Pratt, Michael J. *Computational Geometry for Design and Manufacture*, Ellis Horwood, 1979.
9. Gauss, Karl F. "General Investigations of Curved Surfaces", 1827. Translated by Hiltebeitel, Adam and Morehead, James in *General Investigations of Curved Surfaces*, Raven Press, Hewlett, New York, USA, 1965.
10. Gauss, Karl F. "New General Investigations of Curved Surfaces", 1825. Translated by Hiltebeitel, Adam and Morehead, James in *General Investigations of Curved Surfaces*, Raven Press, Hewlett, New York, USA, 1965.
11. Gordon, William J. "Spline-Blended Surface Interpolation Through Curve Networks", *Journal of Mathematics and Mechanics*, Vol. 18, No. 10, 1969, pp. 931–952. Also Research Publication GMR-921, General Motors Research Laboratories, Warren, Michigan, USA, September 1969.
12. Griss, Martin L. *A REDUCE Symbolic-Numeric Tutorial*, Utah Symbolic Computation Group Operating Note No. UCP-32, Department of Computer Science, University of Utah, Salt Lake City, Utah, USA, October 1977.
13. Hearn, Anthony C. "REDUCE: A User-Oriented Interactive System for Algebraic Simpli- fication". In *Interactive Systems for Experimental Applied Mathematics*, edited by Klerer, M. and Reinfelds, J., Academic Press, New York, USA, 1968.

14. Hearn, Anthony C. *REDUCE 2 User's Manual*, Utah Symbolic Computation Group Report No. UCP-19, Department of Computer Science, University of Utah, Salt Lake City, Utah, USA, March 1973.

15. Hicks, Noel J. *Notes on Differential Geometry*, Van Nostrand, *Mathematical Studies*, Vol. 3, 1965.

16. Lane, Jeffrey M. *Shape Operators for Computer Aided Geometric Design*, Ph.D. Thesis, University of Utah, Salt Lake City, Utah, USA, June 1977.

17. Nielson, Gregory M. "Some Piecewise Polynomial Alternatives to Splines Under Tension". In *Computer Aided Geometric Design*, edited by Barnhill, Robert E. and Riesenfeld, Richard F., Academic Press, New York, USA, 1974, pp. 209–235.

18. Nielson, Gregory M. *Computation of v-splines*, Tech. Report No. 044-433-11, Department of Mathematics, Arizona State University, Tempe, Arizona, USA, June 1974.

19. Pilcher, David T. *Smooth Approximation of Parametric Curves and Surfaces*, Ph.D. Thesis, University of Utah, Salt Lake City, Utah, USA, August 1973.

20. Riesenfeld, Richard F. *Applications of B-spline Approximation to Geometric Problems of Computer-Aided Design*, Ph.D. Thesis, Syracuse University, Syracuse, New York, USA, May 1973. Available as Tech. Report No. UTEC-CSc-73-126, Department of Computer Science, University of Utah, Salt Lake City, Utah, USA.

21. Riesenfeld, Richard F.; Cohen, Elaine; Fish, Russell D.; Thomas, Spencer W.; Cobb, Elizabeth S.; Barsky, Brian A.; Schweitzer, Dino L.; and Lane, Jeffrey M. "Using the Oslo Algorithm as a Basis for CAD/CAM Geometric Modelling." In *Proceedings of the Second Annual NCGA National Conference*, National Computer Graphics Association, Inc., Baltimore, Maryland, USA, 14–18 June 1981, pp. 345–356.

22. Schultz, Martin H. *Spline Analysis*, Prentice-Hall, *Series in Automatic Computation*, Englewood Cliffs, New Jersey, USA, 1973.

23. Schweikert, Daniel G. "An Interpolation Curve Using a Spline in Tension", *Journal of Mathematics and Physics*, Vol. 45, 1966, pp. 312–317.

24. Spivak, Michael, *A Comprehensive Introduction to Differential Geometry, Vol. III*, Publish or Perish, Boston, Massachusetts, USA, 1973.

25. Stoutemyer, David R. "Automatic Error Analysis Using Computer Algebraic Manipulation", *ACM Transactions on Mathematical Software*, Vol. 3, No. 1, March 1977, pp. 26–43.

# Bibliography on Curves and Surfaces

Adams, J. Alan. "A Comparison of Methods for Cubic Spline Curve Fitting", *Computer-Aided Design*, Vol. 6, No. 1, January 1974, pp. 1–9.

Adams, J. Alan. "The Intrinsic Method for Curve Definition", *Computer-Aided Design*, Vol. 7, No. 4, 1975, pp. 243–249. Also Report No. EW-1-75, U.S. Naval Academy, Annapolis, Maryland, USA, 1 January 1975.

Agin, G.J. and Binford, T.O. "Computer Descriptions of Curved Objects", *IEEE Transactions on Computers*, Vol. C-25, No. 4, April 1976, pp. 439–449.

Ahuja, D.V. "An Algorithm for Generating Spline-Like Curves", *IBM Systems Journal*, Vol. 7, No. 3/4, 1968, pp. 206–217.

Akima, Hiroshi. "A New Method of Interpolation and Smooth Curve Fitting Based on Local Procedures", *Journal of the ACM*, Vol. 17, No. 4, October 1970, pp. 589–602.

Akima, Hiroshi. "Algorithm 433. Interpolation and Smooth Curve Fitting Based on Local Procedures", *Communications of the ACM*, Vol. 15, No. 10, October 1972, pp. 914–918.

Akima, Hiroshi. "A Method of Bivariate Interpolation and Smooth Surface Fitting Based on Local Procedures", *Communications of the ACM*, Vol. 17, No. 1, January 1974, pp. 18–20.

Akima, Hiroshi. "Algorithm 474. Bivariate Interpolation and Smooth Surface Fitting Based on Local Procedures", *Communications of the ACM*, Vol. 17, No. 1, January 1974, pp. 26–31.

Akima, Hiroshi. "A Method of Bivariate Interpolation and Smooth Surface Fitting for Irregularly Distributed Data Points", *ACM Transactions on Mathematical Software*, Vol. 4, No. 2, June 1978, pp. 148–159.

Akima, Hiroshi. "Algorithm 526. A Method of Bivariate Interpolation and Smooth Surface Fitting for Irregularly Distributed Data Points", *ACM Transactions on Mathematical Software*, Vol. 4, No. 2, June 1978, pp. 160–164.

Akima, Hiroshi. "Remark on Algorithm 526. A Method of Bivariate Interpolation and Smooth Surface Fitting for Irregularly Distributed Data Points (E1)", *ACM Transactions on Mathematical Software*, Vol. 5, No. 2, June 1979, p. 242.

Akima, Hiroshi. "An Improved Method of Univariate Interpolation Based on Local Procedures", *ACM Transactions on Mathematical Software*.

Akima, Hiroshi. "Algorithm #780712F. Univariate Interpolation Based on Local Procedures", *ACM Transactions on Mathematical Software*.

Anderson, M.R. "Remark on Algorithm 474. Bivariate Interpolation and Smooth Surface Fitting Based on Local Procedures (E2)", *ACM Transactions on Mathematical Software*, Vol. 5, No. 2, June 1979, p. 241.

Armit, Andrew P. *Multipatch Design Program–User's Guide*, Computer-Aided Design Group, Cambridge University, Cambridge, England, September 1968.

Armit, Andrew P. "A Multipatch Design System for Coons' Patches". In *IEEE Conference Publication* No. 51, April 1969.

Armit, Andrew P. *Computer Systems for Interactive Design of Three-Dimensional Shapes*, Ph.D. Thesis, Cambridge University, Cambridge, England, November 1970.

Armit, Andrew P. "Example of an Existing System in University Research. Multipatch and Multiobject Design Systems", *Proceedings of the Royal Society of London*, Vol. A321, 1971, pp. 235–242.

Armit, Andrew P. "Curve and Surface Design Using Multipatch and Multiobject Design Systems", *Computer-Aided Design*, Vol. 3, No. 3, Summer 1971, pp. 3–12.

Armit, Andrew P. "The Interactive Languages of Multipatch and Multiobject Design Systems", *Computer-Aided Design*, Vol. 4, No. 1, Autumn 1971, pp. 10–15.

Armit, Andrew P. "Interactive 3D Shape Design–Multipatch and Multiobject". In *Proceedings of Curved Surfaces in Engineering*, Churchill College, edited by Brown, I.J., IPC Science and Technology Press, Cambridge, England, 15–17 March 1972.

Armit, Andrew P. "A Language for Interactive Shape Design". In *Graphic Languages–Proceedings of the IFIP Working Conference on Graphic Languages*, Vancouver, British Columbia, Canada, edited by Nake, F. and Rosenfeld, A., North-Holland, Amsterdam, 22–26 May 1972, pp. 369–380.

Armit, Andrew P. and Forrest, A. Robin. "Interactive Surface Design". In *Advanced Computer Graphics–Economics, Techniques and Applications*, edited by Parslow, R.D. and Green, R.E., Plenum Press, New York, 1971, pp. 1179–1202.

Artzy, Ehud; Frieder, Gideon; and Herman, Gabor T. "The Theory, Design, Implementation and Evaluation of a Three-Dimensional Surface Detection Algorithm", *Computer Graphics and Image Processing*, Vol. 15, No. 1, January 1981, pp. 1–24. Also in *Proceedings of SIGGRAPH '80*, ACM, Seattle, Washington, USA, July 1980, pp. 2–9.

Ball, A.A. "CONSURF. Part 1: Introduction of the Conic Lofting Tile", *Computer-Aided Design*, Vol. 6, No. 4, 1974, pp. 243–249.

Ball, A.A. "CONSURF. Part 2: Description of the Algorithms", *Computer-Aided Design*, Vol. 7, No. 4, 1975, pp. 237–242.

Ball, A.A. "CONSURF. Part 3: How the Program is Used", *Computer-Aided Design*, Vol. 9, No. 1, January 1977, pp. 9–12.

Ball, A.A. "Simple Specification of the Parametric Cubic Segment", *Computer-Aided Design*, Vol. 10, No. 3, May 1978, pp. 181–182.

Bandurski, A.E. and Jefferson, D. "Enhancements to the DBTG Model for Computer-Aided Ship Design". In *Proceedings of the Workshop on Databases for Interactive Design*, University of Waterloo, Waterloo, Ontario, Canada, 15–16 September 1975.

Bar, G. "Parametrische Interpolation empirischer Raumkurven", *ZAMM*, Vol. 57, 1977, pp. 305–314.

Barnhill, Robert E. "Smooth Interpolation over Triangles". In *Computer Aided Geometric Design*, edited by Barnhill, Robert E. and Riesenfeld, Richard F., Academic Press, New York, USA, 1974, pp. 45–70.

Barnhill, Robert E. "Blending Function Interpolation: A Survey and Some New Results". In *Proceedings of the Conference on Numerical Methods in Approximation Theory*. Oberwolfach, Germany, ISNM 30, 1975, pp. 43–90. Also Numerical Analysis Report No. 9, University of Dundee, Scotland, July 1975.

Barnhill, Robert E. "Blending Function Finite Elements for Curved Boundaries". In *Finite Elements and Their Applications*, edited by Whiteman, J.R., Academic Press, New York, USA, 1976, pp. 67–76.

Barnhill, Robert E. "Representation and Approximation of Surfaces". In *Mathematical Software III*, edited by Rice, John R., Academic Press, New York, USA, 1977, pp. 68–119.

Barnhill, Robert E.; Birkhoff, Garrett; and Gordon, William J. "Smooth Interpolation in Triangles", *Journal of Approximation Theory*, Vol. 8, 1973, pp. 114–128.

Barnhill, Robert E. and Brown, James H. *Curved Nonconforming Elements for Plate Problems*, Numerical Analysis Report No. 8, University of Dundee, Scotland, August 1975.

Barnhill, Robert E.; Brown, James H.; and Klucewicz, I. Michele. "A New Twist for Computer Aided Geometric Design", *Computer Graphics and Image Processing*, Vol. 8, No. 1, August 1978, pp. 78–91.

Barnhill, Robert E.; Dube, R. Peter; and Little, Frank F. *Surfaces of Interpolation to Arbitrarily Spaced Data*, University of Utah, Salt Lake City, Utah, USA, 1978.

Barnhill, Robert E. and Gregory, John A. "Polynomial Interpolation to Boundary Data on

Triangles", *Mathematics of Computation*, Vol. 29, 1975, pp. 726–735.

Barnhill, Robert E. and Gregory, John A. "Compatible Smooth Interpolation in Triangles", *Journal of Approximation Theory*, Vol. 15, No. 3, 1975, pp. 214–225.

Barnhill, Robert E. and Mansfield, Lois. "Error Bounds for Smooth Interpolation on Triangles", *Journal of Approximation Theory*, Vol. 11, No. 4, 1974, pp. 306–318.

Barnhill, Robert E. and Riesenfeld, Richard F. (editors). *Computer Aided Geometric Design*, Academic Press, New York, USA, 1974.

Barnhill, Robert E. and Riesenfeld, Richard F. "Recent Developments in Computer Aided Geometric Design". In *Proceedings of SNAME, 1st International Symposium on Computer Aided Hull Surface Definition*, Annapolis Conference, Annapolis, Maryland, USA, 1977, pp. 97–103.

Barnhill, Robert E. and Riesenfeld, Richard F. "Surface Representation for Computer Aided Design". In *Data Structures, Computer Graphics, and Pattern Recognition*, edited by Klinger, A.; Fu, K.S.; and Kunii, T.L., Academic Press, New York, USA, 1977, pp. 413–426.

Barr, Alan H. "Superquadrics and Angle-Preserving Transformations", *IEEE Computer Graphics and Applications*, Vol. 1, No. 1, January 1981, pp. 11–23.

Barsky, Brian A. *A Method for Describing Curved Surfaces by Transforming Between Interpolatory Spline and B-spline Representations*, Master's Thesis, Cornell University, Ithaca, New York, USA, January 1979.

Barsky, Brian A. "Computer-Aided Geometric Design: A Bibliography with Keywords and Classified Index", *IEEE Computer Graphics and Applications*, Vol. 1, No. 3, July 1981, pp. 67–109. Also reprinted in *ACM Computer Graphics*, Vol. 16, No. 1, May 1982, pp. 119–159.

Barsky, Brian A. *The Beta-spline: A Local Representation Based on Shape Parameters and Fundamental Geometric Measures*, Ph.D. Thesis, University of Utah, Salt Lake City, Utah, USA, December 1981.

Barsky, Brian A. "End Conditions and Boundary Conditions for Uniform B-spline Curve and Surface Representations", *Computers in Industry*, Vol. 3, Nos. 1 and 2, March and June 1982, pp. 17–29. Special Steven A. Coons Memorial Issue.

Barsky, Brian A. *A Study of the Parametric Uniform B-spline Curve and Surface Representations*, Tech. Report No. UCB/CSD 83/118, Computer Science Division, Electrical Engineering and Computer Sciences Department, University of California, Berkeley, California, USA, May 1983.

Barsky, Brian A. "A Description and Evaluation of Various 3-D Models", *IEEE Computer Graphics and Applications*, Vol. 4, No. 1, January 1984, pp. 38–52. Earlier version published in *Proceedings of InterGraphics '83*, Japan Management Association, Tokyo, 11–14 April 1983, pp. (B2–5) 1 to 21 and reprinted in *Computer Graphics – Theory and Applications*, edited by Kunii, Tosiyasu L., Springer-Verlag, Berlin, Germany, 1983, pp. 75–95.

Barsky, Brian A. "Exponential and Polynomial Methods for Applying Tension to an Interpolating Spline Curve", *Computer Vision, Graphics, and Image Processing*, Vol. 27, No. 1, July 1984, pp. 1–18.

Barsky, Brian A. "An Explanation of the Beta-spline". In *Proceedings of Computer Graphics '85*, Prague, Czechoslovakia, 26–28 March 1985, pp. 15–33.

Barsky, Brian A. *Arbitrary Subdivision of Bézier Curves*, Tech. Report No. UCB/CSD 85/265, Computer Science Division, Electrical Engineering and Computer Sciences Department, University of California, Berkeley, California, USA, October 1985.

Barsky, Brian A. "The Beta-spline: A Curve and Surface Representation for Computer Graphics and Computer Aided Geometric Design". In *Techniques in Computer Graphics*, Proceedings of the International Summer Institute, Stirling, Scotland, 29 June–4 July 1986, edited by Rogers, David F. and Earnshaw, Rae A., Springer-Verlag, Heidelberg, 1987, p. 65.

Barsky, Brian A. and Beatty, John C. *Varying the Betas in Beta-splines*, Tech. Report No. UCB/CSD 82/112, Computer Science Division, Electrical Engineering and Computer Sciences Department, University of California, Berkeley, California, USA, December 1982. Also Tech. Report No. CS-82-49, Department of Computer Science, University of Waterloo, Waterloo, Ontario, Canada.

Barsky, Brian A. and Beatty, John C. "Controlling the Shape of Parametric B-spline and Beta-spline Curves". In *Proceedings of Graphics Interface '83*, Edmonton, Alberta, Canada, 8–13 May 1983, pp. 223–232.

Barsky, Brian A. and Beatty, John C. "Local Control of Bias and Tension in Beta-splines", *ACM Transactions on Graphics*, Vol. 2, No. 2, April 1983, pp. 109–134. Also published in *Proceedings of SIGGRAPH '83*, Vol. 17, No. 3, ACM, Detroit, Michigan, USA, 25–29 July 1983, pp. 193–218.

Barsky, Brian A. and DeRose, Tony D. *Geometric Continuity of Parametric Curves*, Tech. Report No. UCB/CSD 84/205, Computer Science Division, Electrical Engineering and Computer Sciences Department, University of California, Berkeley, California, USA, October 1984.

Barsky, Brian A. and DeRose, Tony D. "The Beta2-spline: A Special Case of the Beta-spline Curve and Surface Representation", *IEEE Computer Graphics and Applications*, Vol. 5, No. 9, September 1985, pp. 46–58. Correction published in Letter to the Editor, *IEEE Computer Graphics and Applications*, Vol. 7, No. 3, March 1987, p. 15. Earlier version published as Tech. Report No. UCB/CSD 83/152, Computer Science Division, Electrical Engineering and Computer Sciences Department, University of California, Berkeley, California, USA, November 1983.

Barsky, Brian A.; DeRose, Tony D.; and Dippé, Mark D. "An Adaptive Subdivision Method with Crack Prevention for Rendering Beta-spline Objects", Tech. Report No. UCB/CSD 87/348, Computer Science Division, Electrical Engineering and Computer Sciences Department, University of California, Berkeley, California, USA, March 1987.

Barsky, Brian A. and Fournier, Alain. "Computational Techniques for Parametric Curves and Surfaces". In *Proceedings of Graphics Interface '82*, Canadian Man-Computer Communications Society and National Computer Graphics Association of Canada, Toronto, Ontario, Canada, 17–21 May 1982, pp. 57–71.

Barsky, Brian A. and Greenberg, Donald P. "Determining a Set of B-spline Control Vertices to Generate an Interpolating Surface", *Computer Graphics and Image Processing*, Vol. 14, No. 3, November 1980, pp. 203–226.

Barsky, Brian A. and Greenberg, Donald P. "An Interactive Surface Representation System Using a B-spline Formulation with Interpolation Capability", *Computer-Aided Design*, Vol. 14, No. 4, July 1982, pp. 187–194. Corrigendum in *Computer-Aided Design*, Vol. 15, No. 3, May 1983, p. 174.

Barsky, Brian A. and Thomas, Spencer W. *TRANSPLINE Curve Representation System*, Tech. Report No. UUCS-80-104, Department of Computer Science, University of Utah, Salt Lake City, Utah, USA, April 1980.

Barsky, Brian A. and Thomas, Spencer W. "TRANSPLINE–A System for Representing Curves Using Transformations Among Four Spline Formulations", *The Computer Journal*, Vol. 24, No. 3, August 1981, pp. 271–277.

Bartels, Richard H. and Beatty, John C. *Beta-splines with a Difference*, Tech. Report No. CS-83-40, Department of Computer Science, University of Waterloo, Waterloo, Ontario, Canada, May 1984.

Bartels, Richard H.; Beatty, John C.; and Barsky, Brian A. *An Introduction to Splines for Use in Computer Graphics and Geometric Modeling*, Morgan Kaufmann Publishers, Inc., Los Altos, California, 1987.

Bates, K.J. "The AUTOKON AUTOMOTIVE and AEROSPACE Packages". In *Proceedings of Curved Surfaces in Engineering*, Churchill College, edited by Brown, I.J., IPC Science and Technology Press, Cambridge, England, 15–17 March 1972.

Belser, K. "Comment on 'An Improved Algorithm for the Generation of Nonparametric Curves'", *IEEE Transactions on Computers*, Vol. C-25, No. 1, January 1976, p. 103.

Bézier, Pierre E. "How Renault Uses Numerical Control for Car Body Design and Tooling, SAE Paper 680010". In *Society of Automotive Engineers Congress*, Detroit, Michigan, USA, 1968.

Bézier, Pierre E. "Procédé de définition numérique des courbes et surfaces non mathématiques; Système UNISURF", *Automatisme*, Vol. 13, May 1968.

Bézier, Pierre E. *Procédé UNISURF de définition numérique des surfaces non mathématiques*, Tech. Report No. 219, Mécanique Electricité, March 1968.

Bézier, Pierre E. *Emploi des machines à commande numérique*, Masson, Paris, France, 1970. Translated by Forrest, A. Robin and Pankhurst, Anne F. as *Numerical Control–Mathematics and Applications*, John Wiley, London, 1972.

Bézier, Pierre E. "Example of an Existing System in the Motor Industry: The UNISURF System", *Proceedings of the Royal Society of London*, Vol. A321, 1971, pp. 207–218.

Bézier, Pierre E. "UNISURF System: Principles, Program, Language". In *Proceedings of the 1973 International Conference on Programming Languages for Numerically Controlled Machine Tools*, Budapest, Hungary, edited by Hatvany, Joseph, North-Holland, Amsterdam, Holland, April 1973.

Bézier, Pierre E. "Mathematical and Practical Possibilities of UNISURF". In *Computer Aided Geometric Design*, edited by Barnhill, Robert E. and Riesenfeld, Richard F., Academic Press, New York, USA, 1974.

Bézier, Pierre E. *Essai de définition numérique des courbes et des surfaces expérimentales*, Ph.D. Thesis, Université Pierre et Marie Curie, Paris, France, February 1977.

Bézier, Pierre E. "General Distortion of an Ensemble of Biparametric Surfaces", *Computer-Aided Design*, Vol. 10, No. 2, March 1978, pp. 116–120.

Birkhoff, G. and Garabedian, H.L. "Smooth Surface Interpolation", *Journal of Mathematics and Physics*, Vol. 39. 1960, pp. 258–268.

Birkhoff, Garrett and Gordon, William J. "The Draftsman's and Related Equations", *Journal of Approximation Theory*, Vol. 1, No. 2, 1968, pp. 199–208.

Blinn, James F. "Models of Light Reflection for Computer Synthesized Pictures". In *Proceedings of SIGGRAPH '77*, ACM, San José, California, USA, 20–22 July 1977, pp. 192–198.

Blinn, James F. *Computer Display of Curved Surfaces*, Ph.D. Thesis, University of Utah, Salt Lake City, Utah, USA, December 1978. Also Tech. Report No. 1060–126, Jet Propulsion Laboratory, Pasadena, California, USA.

Blinn, James F. "Simulation of Wrinkled Surfaces". In *Proceedings of SIGGRAPH '78*, ACM, Atlanta, Georgia, USA, 23–25 August 1978, pp. 286–292.

Blinn, James F. and Newell, Martin E. "Texture and Reflection in Computer Generated Images", *Communications of the ACM*, Vol. 19, No. 10, October 1976, pp. 542–547.

Bloor, M.S.; de Pennington, A.; and Woodwark, J.R. "RISP: Bridging the Gap Between Conventional Surface Elements". In *Proceedings of the CAD78 Conference*, Brighton, England, IPC Science and Technology Press, Guildford, England, 1978.

Boehm, Wolfgang. "Parameterdarstellung kubischer und bikubischer Splines", *Computing*, Vol. 17, 1976, pp. 87–92.

Boehm, Wolfgang. "Über die Konstruktion von B-spline-Kurven", *Computing*, Vol. 18, No. 2, 1977, pp. 161–167.

Boehm, Wolfgang. "Cubic B-spline Curves and Surfaces in Computer-Aided Geometric Design", *Computing*, Vol. 19, No. 1, 1977, pp. 29–34.

Boehm, Wolfgang. "Inserting New Knots into B-spline Curves", *Computer-Aided Design*, Vol. 12, No. 4, July 1980, pp. 199–201.

Boere, H.: *The Design of Surfaces with the Coons' Method*, Ph.D. Thesis, Technical University of Delft, The Netherlands, November 1970.

Bolton, K.M. "Biarc Curves", *Computer-Aided Design*, Vol. 7, No. 2, 1975, pp. 89–92.

Bresenham, J.E. "A Linear Algorithm for the Incremental Digital Display of Circular Arcs", *Communications of the ACM*, Vol. 20, No. 2, February 1977, pp. 100–106.

Brewer, John A. and Anderson, David C. "Visual Interaction with Overhauser Curves and Surfaces". In *Proceedings of SIGGRAPH '77*, ACM, San José, California, USA, 20–22 July 1977, pp. 132–137.

Brodlie, K.W. (editor). *Mathematical Methods in Computer Graphics and Design*, Academic Press, New York, USA, 1979.

Brown, Bruce E. *Modeling of Solids for Three-Dimensional Finite Element Analysis*, Ph.D. Thesis, University of Utah, Salt Lake City, Utah, USA, June 1974.

Brown, James H. *Conforming and Nonconforming Finite Element Models for Curved Regions*, Ph.D. Thesis, University of Dundee, Scotland, June 1976.

Brueckner, I. "Construction of Bezier Points of Quadrilaterals from Those of Triangles", *Computer-Aided Design*, Vol. 12, No. 1, January 1980, pp. 21–24.

Calu, Johan V.J. *Implementation of Surface Interpolation for Computer Graphics*, Master's Thesis, University of Utah, Salt Lake City, Utah, USA, December 1974.

Carpenter, Loren C. "Computer Rendering of Fractal Curves and Surfaces", *Communications of* the ACM. In *Proceedings of SIGGRAPH '80*, Special Supplement, ACM, Seattle, Washington, USA, July 1980.

Catmull, Edwin E. *A Subdivision Algorithm for Computer Display of Curved Surfaces*, Ph.D. Thesis, University of Utah, Salt Lake City, Utah, USA, December 1974. Also Tech. Report No. UTEC-CSc-74-133, Department of Computer Science, University of Utah.

Catmull, Edwin E. "Computer Display of Curved Surfaces". In *Proceedings of the IEEE Conference on Computer Graphics, Pattern Recognition, and Data Structure*, Los Angeles, California, USA, 14–16 May 1975, pp. 11–17.

Catmull, Edwin E. and Clark, James H. "Recursively Generated B-spline Surfaces on Arbitrary Topological Meshes", *Computer-Aided Design*, Vol. 10, No. 6, November 1978, pp. 350–355.

Catmull, Edwin E. and Rom, Raphael J. "A Class of Local Interpolating Splines". In *Computer Aided Geometric Design*, edited by Barnhill, Robert E. and Riesenfeld, Richard F., Academic Press, New York, USA, 1974, pp. 317–326.

Cerny, H.F. *F-curve and F-surface Layout*, Document No. D2-23924-1, Boeing Company, Seattle, Washington, USA, March 1965.

Chaikin, George M. "An Algorithm for High-Speed Curve Generation", *Computer Graphics and Image Processing*, Vol. 3, 1974, pp. 346–349.

Cheng, Kuo-Young; Chen, Wen-Tsuen; and Li Sy-Tey. "An Approach to Approximation of Curves for Shipbuilding Applications", *J. Chin, Inst. Eng. Taiwan*, Vol. 2, No. 1, January 1979, pp. 23–29.

Ciaffi, F. and Valle, G. "Interactive Surface Molding for Car Body Design". In *Computer Graphics Symposium*, Berlin, Germany, 1971, pp. 255–274.

Clark, James H. *3-D Design of Free-Form B-spline Surfaces*, Ph.D. Thesis, University of Utah, Salt Lake City, Utah, USA, September 1974. Also Tech. Report No. UTEC-CSc-74-120, Department of Computer Science, University of Utah.

Clark, James H. "Some Properties of B-splines". In *Proceedings of the 2nd USA-Japan Computer Conference*, AFIPS, Montvale, New Jersey, USA, 26–28 August 1975, pp. 542–545.

Clark, James H. "Designing Surfaces in 3-D", *Communications of the ACM*, Vol. 19, No. 8, August 1976, pp. 454–460.

Clark, James H. "Hierarchical Geometric Models for Visible Surface Algorithms", *Communications of the ACM*, Vol. 19, No. 10, October 1976, pp. 547–554.

Clark, James H. "A Fast Scan-Line Algorithm for Rendering Parametric Surfaces", *Communications of the ACM*. In *Proceedings of SIGGRAPH '79*, Special Supplement, ACM, Chicago, Illinois, USA, 8–10 August 1979.

Cline, Alan K. "Scalar- and Planar-Valued Curve Fitting Using Splines Under Tension", *Communications of the ACM*, Vol. 17, No. 4, April 1974, pp. 218–220. Also in *Atmos. Tech.*, No. 3, 1973, pp. 60–65.

Cohen, Dan. "On Linear Difference Curves". In *Advanced Computer Graphics – Economics, Techniques and Applications*, edited by Parslow, R.D. and Green, R.E., Plenum Press, New York and London, 1971, pp. 1143–1177.

Cohen, D. and Lee, T.M.P. "Fast Drawing of Curves for Computer Display". In *Proceedings of the Spring Joint Computer Conference*, AFIPS Press, Montvale, New Jersey, USA, 1969, pp. 297–307.

Cohen, Elaine; Lyche, Tom; and Riesenfeld, Richard F. "Discrete B-splines and Subdivision Techniques in Computer-Aided Geometric Design and Computer Graphics", *Computer Graphics and Image Processing*, Vol. 14, No. 2, October 1980, pp. 87–111. Also Tech. Report No. UUCS-79-117, Department of Computer Science, University of Utah, Salt Lake City, Utah, USA, October 1979.

Cohen, Elaine and Riesenfeld, Richard F. "An Incompatibility Projector Based on an Interpolant of Gregory", *Computer Graphics and Image Processing*, Vol. 8, No. 2, 1978, pp. 294–298.

Cohen, Elaine and Riesenfeld, Richard F. "General Matrix Representations for Bézier and B-spline Curves", *Computers in Industry*, Vol. 3, Nos. 1 and 2, March and June 1982. Special Steven A. Coons Memorial Issue.

Coles, W. A. "Use of Graphics in an Aircraft Design Office", *Computer-Aided Design*, Vol. 9, 1977, pp. 23–28.

Collins, P.S. and Gould, S.S. "Computer-Aided Design and Manufacture of Surfaces for Bottle

Moulds". In *Proceedings of the CAM74 Conference on Computer Aided Manufacture and Numerical Control*, Strathclyde University, 1974.

Computer Aided Manufacturing International, Inc. *User Documentation for Sculptured Surfaces Releases SSX5 and SSX5A*, Publication No. PS-76-SS-02, Arlington, Texas, USA, 1976.

Computer Aided Manufacturing International, Inc. *Sculptured Surfaces Users Course*, Publication No. TM-77-SS-01, Arlington, Texas, USA, 1977.

Coons, Steven A. "An Outline of the Requirements for a Computer-Aided Design System". In *Proceedings of the Spring Joint Computer Conference*, AFIPS, Spartan Books, Washington, DC, USA, 1963, pp. 299–304. Also *Simulation*, Vol. II, Issue 2, p. 64.

Coons, Steven A. *Surfaces for Computer Aided Design*, Design Division, Mechanical Engineering Department, MIT, Cambridge, Massachusetts, USA, 1964.

Coons, Steven A. *Surfaces for Computer-Aided Design of Space Forms*, Tech. Report No. MAC-TR-41, Project MAC, MIT, Cambridge, Massachusetts, USA, June 1967. Available as AD-663 504 from NTIS, Springfield, Virginia, USA.

Coons, Steven A. *Rational Bicubic Surfaces Patches*, Project MAC, MIT, Cambridge, Massachusetts, USA, November 1968.

Coons, Steven A. "Surface Patches and B-spline Curves". In *Computer Aided Geometric Design*, edited by Barnhill, Robert E. and Riesenfeld, Richard F., Academic Press, New York, USA, 1974, pp. 1–16.

Coons, Steven A. "Modification of the Shape of Piecewise Curves", *Computer-Aided Design*, Vol. 9, No. 3, July 1977, pp. 178–180.

Coons, Steven A. "Constrained Least Squares", *Computers and Graphics*, Vol. 3, 1978, pp. 43–47.

Coons, Steven A. and Herzog, Bertram. "Surfaces for Computer-Aided Aircraft Design, Paper 67-895". In *American Institute of Aeronautics and Astronautics*, 1967.

Cox, Morris G. and Hayes, Jeffrey G. *Curve Fitting: A Guide and Suite of Programs for the Nonspecialist User*. Report No. NPL-DNACS-26, National Physical Laboratory, Teddington, Middlesex, England, 1973.

Creutz, Guenter and Schubert, Christian. "An Interactive Line Creation Method Using B-splines". *Computers and Graphics*, Vol. 5, No. 2–4, 1980, pp. 69–78.

Danielsson, Per-Erik. "Incremental Curve Generation", *IEEE Transactions on Computers*, Vol. C-19, No. 9, September 1970, pp. 783–793.

Danielsson, Per-Erik. "Comments on Circle Generator for Display Devices", *Computer Graphics and Image Processing*, Vol. 7, No. 2, April 1978, pp. 300–301.

DATASAAB. *FORMELA General Description*, Reg. No. 917-E, SAAB Aktiebolag 58188, Linkoping, Sweden, 1965.

DeRose, Tony D. and Barsky, Brian A. "Geometric Continuity and Shape Parameters for Catmull-Rom Splines (Extended Abstract)". In *Proceedings of Graphics Interface '84*, Ottawa, Ontario, Canada, 27 May–1 June 1984, pp. 57–64.

DeRose, Tony D. and Barsky, Brian A. "An Intuitive Approach to Geometric Continuity for Parametric Curves and Surfaces". In *Proceedings of Graphics Interface '85*, Montreal, Quebec, Canada, 27–31 May 1985, pp. 343–351. Extended abstract in *Proceedings of the International Conference on Computational Geometry and Computer-Aided Design*, New Orleans, Louisiana, USA, 5–8 June 1985, pp. 71–75. Revised version published in *Computer-Generated Images—The State of the Art*, edited by Magnenat-Thalmann, Nadia and Thalmann, Daniel, Springer-Verlag, Heidelberg, 1985, pp. 159–175.

DeRose, Tony D. and Barsky, Brian A. "Geometric Continuity, Shape Parameters, and Geometric Constructions for Catmull-Rom Splines", *ACM Transactions on Graphics*, to appear.

Dimsdale, B. "Bicubic Patch Bounds", *Comp. & Maths. with Appls.*, Vol. 3, No. 2, 1977, pp. 95–104.

Dimsdale, B. and Burkley, R.M. "Bicubic Patch Surfaces for High-Speed Numerical Control Processing", *IBM J. Res. and Develop.*, 1976, pp. 358–367.

Dimsdale, B. and Johnson, K. "Multiconic Surfaces", *IBM J. Res. and Develop.*, Vol. 19, No. 6, 1975, pp. 523–529.

Dokken, Tor and Ulfsby, Stig. *Specifications of the Modules for Sculptured Surfaces*, GPM Report No. 5, Central Institute for Industrial Research, 1979.

Dollries, J.F. *Three-Dimensional Surface Fit and Numerically Controlled Machining from a Mesh of Points*, Technical Information Series Report No. R63FPD319, General Electric Company, 1963.

Done, G.T.S. "Interpolation of Mode Shapes: A Matrix Scheme Using Two-Way Spline Curves", *The Aeronautical Quarterly*, November 1965, pp. 333–349.

Doo, D.W.H. "A Subdivision Algorithm for Smoothing Down Irregular Shaped Polyhedrons". In *Proceedings of the International Conference Interactive Techniques in Computer Aided Design*, IEEE Computer Society, Bologna, Italy, 21–23 September 1978, pp. 157–165.

Doo, D.W.H. and Sabin, M.A. "Behaviour of Recursive Division Surfaces Near Extraordinary Points", *Computer-Aided Design*, Vol. 10, No. 6, November 1978, pp. 356–360.

Doran, C. "Remark on Algorithm 475 (J6)–Visible Surface Plotting Programs", *Communications of the ACM*, Vol. 18, No. 5, May 1975, pp. 277.

Du, Wen-Hui; Schmitt, Francis J.M.; and Barsky, Brian A. "Modelling Free-form Surfaces Using Brown's Interpolant with Control Parameters", in *Proceedings of the International Conference on Computer-Aided Drafting, Design, and Manufacturing Technology*, pp. 240–247, Beijing, 21–25 April 1987.

Dube, R. Peter. *Local Schemes for Computer Aided Geometric Design*, Ph.D. Thesis, University of Utah, Salt Lake City, Utah, USA, June 1975.

Dube, R. Peter. "Tension in a Bicubic Surface Patch", *Computer Graphics and Image Processing*, Vol. 5, December 1976, pp. 496–502.

Dube, R. Peter. "Univariate Blending Functions and Alternatives", *Computer Graphics and Image Processing*, Vol. 6, 1977, pp. 394–408.

Dube, R. Peter. "Preliminary Specification of Spline Curves", *IEEE Transactions on Computers*, Vol. C-28, No. 4, April 1979, pp. 286–290.

Dube, R. Peter; Herron, Gary J.; Little, Frank F.; and Riesenfeld, Richard F. "SURFED-An Interactive Editor for Free-Form Surfaces", *Computer-Aided Design*, Vol. 10, No. 2, March 1978, pp. 111–115.

Duncan, J.M. *Application of Differential Geometry to Computer Curves and Surfaces*, Ph.D. Thesis, University of Durham, Durham, England, 1976.

Duncan, J.P. and Vickers, G.W. "Simplified Method for Interactive Adjustment of Surfaces". *Computer-Aided Design*, Vol. 12, No. 6, November 1980, pp. 305–308.

Einar, H. and Skappel, E. "FORMELA: A General Design and Production Data System for Sculptured Products", *Computer-Aided Design*, Vol. 5, No. 2, 1973, pp. 68–76.

Ellis, T.M.R. and McLain, D.H. "Algorithm 514. A New Method of Cubic Curve Fitting Using Local Data (E2)", *ACM Transactions on Mathematical Software*, Vol. 3, No. 2, June 1977, pp. 175–178.

Emmerson, W.C. "CAD in the Motor Inductry", *Computer-Aided Design*, Vol. 8, No. 3, 1976, pp. 193–197.

England, J.N. "A System for Interactive Modeling of Physical Curved Surface Objects". In *Proceedings of SIGGRAPH '78*, ACM, Atlanta, Georgia, USA, 23–25 August 1978, pp. 336–340.

Ensign, M.G. *A Polynomial Boolean Sum Interpolant for Computer Aided Geometric Design*, Master's Thesis, University of Utah, Salt Lake City, Utah, USA, 1976.

Epstein, M.P. "On the Influence of Parametrisation in Parametric Interpolation", *SIAM Journal on Numerical Analysis*, Vol. 13, No. 2, 1976, pp. 261–268.

Farin, Gerald. *Subsplines über Dreiecken*, Ph.D. Thesis, Technische Universität Braunschweig, Braunschweig, West Germany, 1979.

Faux, Ivor D. and Pratt, Michael J. *Computational Geometry for Design and Manufacture*, Ellis Horwood, 1979.

Feng, David Y. *A Symbolic System for Computer-Aided Development of Surface Interpolants*, Master's Thesis, University of Utah, Salt Lake City, Utah, USA, 1977.

Feng, David Y. and Riesenfeld, Richard F. "A Symbolic System for Computer-Aided Development of Surface Interpolants", *Software–Practice and Experience*, Vol. 8, No. 4, July–August 1978, pp. 461–481.

Feng, David Y. and Riesenfeld, Richard F. "Some New Surface Forms for Computer-Aided Geometric Design", *The Computer Journal*, Vol. 23, No. 4, November 1980, pp. 324–331.

Ferguson, James C. "Multivariable Curve Interpolation", *Journal of the ACM*, Vol. 11, No. 2, April 1964, pp. 221–228. Also Report No. D2-22504, The Boeing Company, Seattle, Washington, USA, 1963.

Flanagan, D.L. and Hefner, O.V. "Surface Moulding–New Tool for the Engineer", *Aeronautics and Astronautics*, April 1967, pp. 58–62.

Flutter, A.G. "The POLYSURF System". In *Proceedings of the 1973 International Conference on Programming Languages for Numerically Controlled Machine Tools*, Budapest, Hungary, edited by Hatvany, Joseph, North-Holland, Amsterdam, Holland, April 1973.

Flutter, A.G. and Rolph, R.N. "POLYSURF: An Interactive System for Computer-Aided Design and Manufacture of Components". In *Proceedings of the CAD76 Conference*, IPC Science and Technology Press, London, England, 1976, pp. 150–158.

Forrest, A. Robin. *Curves and Surfaces for Computer-Aided Design*, Ph.D. Thesis, Cambridge University, Cambridge, England, July 1968.

Forrest, A. Robin, "Curves for Computer Graphics". In *Pertinent Concepts in Computer Graphics*, University of Illinois Press, 1969.

Forrest, A. Robin. *The Computation of Bicubic Twist Terms*, CAD Group Document No. 20, Computer Aided Design Group, Cambridge University, Cambridge, England, January 1969.

Forrest, R. Robin. *The Twisted Cubic Curve*, CAD Group Document No. 50, Cambridge University, Cambridge, England, November 1970.

Forrest, A. Robin. "Definition des surfaces", *Ingenieurs de l'automobile*, Vol. 44, No. 10, October 1971, pp. 521–527.

Forrest, A. Robin. "Interactive Interpolation and Approximation by Bézier Polynomials", *The Computer Journal*, Vol. 15, No. 1, January 1972, pp. 71–79. Also CAD Group Document No. 45, UML, Cambridge University, Cambridge, England, October 1970.

Forrest, A. Robin. "Mathematical Principles for Curve and Surface Representation". In *Proceedings of Curved Surfaces in Engineering*, Churchill College, edited by Brown, I.J., IPC Science and Technology Press, Cambridge, England, 15–17 March 1972, pp. 5–13.

Forrest, A. Robin. "On Coons' and Other Methods for the Representation of Curved Surfaces", *Computer Graphics and Image Processing*, Vol. 1, No. 4, December 1972, pp. 341–359.

Forrest, A. Robin. "Computational Geometry–Achievements and Problems". In *Computer Aided Geometric Design*, edited by Barnhill, Robert E. and Riesenfeld, Richard F., Academic Press, New York, USA, 1974.

Forrest, A. Robin. "A Unified Approach to Geometric Modelling". In *Proceedings of SIGGRAPH '78*, ACM, Atlanta, Georgia, USA, 23–25 August 1978, pp. 264–269.

Forrest, A. Robin. "Research Trends in Computer-Aided Geometric Design". In *Proceedings of the International Conference Interactive Techniques in Computer Aided Design*, IEEE Computer Society, Bologna, Italy, 21–23 September 1978, pp. 141–146.

Forrest, A. Robin. "On the Rendering of Surfaces". In *Proceedings of SIGGRAPH '79*, ACM, Chicago, Illinois, USA, 8–10 August 1979, pp. 253–257. Also published in March 1979 as Tech. Report No. UUCS-79-104, Department of Computer Science, University of Utah, Salt Lake City, Utah, USA, and as Memo CGP 79/1, School of Computing Studies and Accountancy, Computational Geometry Project, University of East Anglia, Norwich, England.

Forrest, A. Robin. "New Dimensions in Spatial Graphics". In *Infotech State of the Art Conference on Computer Graphics*, London, England, 29–31 October 1979. Also Memo CGP 79/8, School of Computing Studies and Accountancy, Computational Geometry Project, University of East Anglia, Norwich, England, September 1979.

Forrest, A. Robin "The Twisted Cubic Curve: A Computer-Aided Geometric Design Approach", *Computer-Aided Design*, Vol. 12, No. 4, July 1980, pp. 165–172. Also Memo CGP 79/2, School of Computing Studies and Accountancy, Computational Geometry Project, University of East Anglia, Norwich, England, July 1979.

Fournier, Alain and Barsky, Brian A. "Geometric Continuity with Interpolating Bézier Curves (Extended Summary)". In *Proceedings of Graphics Interface '85*, Montreal, Quebec, Canada, 27–31 May 1985, pp. 337–341. Revised version published in *Computer-Generated Images—The State of the Art*, edited by Magnenat-Thalmann, Nadia and Thalmann, Daniel, Springer-Verlag, Heidelberg, 1985, pp. 11–25.

Fournier, Alain and Fussell, Don. "Stochastic Modeling in Computer Graphics", *Communica-*

*tions of the ACM*. In *Proceedings of SIGGRAPH '80*, Special Supplement, ACM, Seattle, Washington, 14–18 July 1980.

Frank, Amalie J. "Parametric Font and Image Definition and Generation". In *Proceedings of the Fall Joint Computer Conference*, AFIPS Press, Montvale, New Jersey, USA, 1971, pp. 135–144.

Franke, R.H. "Locally Determined Smooth Interpolation at Irregularly Spaced Points in Several Variables", *J. Inst. Maths. Applics.*, Vol. 19, 1977, pp. 471–482. Also Naval Postgraduate School Technical Report.

Franke, R.H. *Smooth Surface Approximation by a Local Method of Interpolation at Scattered Points*, Tech. Report No. NPS53-78-002, Naval Postgraduate School, 1978.

Freemantle, A.C. and Freeman, P.L. "The Evolution and Application of Lofting Techniques at Hawker Siddeley Aviation". In *Proceedings of Curved Surfaces in Engineering*, Churchill College, edited by Brown, I.J., IPC Science and Technology Press, Cambridge, England, 15–17 March 1972.

Fuchs, Henry; Kedem, Zvi M.; and Uselton, Sam P. "Optimal Surface Reconstruction from Planar Contours", *Communications of the ACM*, Vol. 20, No. 10, October 1977, pp. 693–702.

Gauss, Karl F. "General Investigations of Curved Surfaces", 1827. Translated by Hiltebeitel, Adam and Morehead, James in *General Investigations of Curved Surfaces*, Raven Press, Hewlett, New York, USA, 1965.

Gauss, Karl F. "New General Investigations of Curved Surfaces", 1825. Translated by Hiltebeitel, Adam and Morehead, James in *General Investigations of Curved Surfaces*, Raven Press, Hewlett, New York, USA, 1965.

Ghezzi, C. and Tisato, F. "Interactive Computer-Aided Design for Sculptured Surfaces". In *Proceedings of the 1973 International Conference on Programming Languages for Numerically Controlled Machine Tools*, Budapest, Hungary, edited by Hatvany, Joseph, North-Holland, Amsterdam, Holland, April 1973.

Gill, J.I. *Computer-Aided Design of Shell Structures Using the Finite Element Method*, Ph.D. Thesis, Cambridge University, Cambridge, England, 1972.

Godwin, A.N. "Family of Cubic Splines with One Degree of Freedom", *Computer-Aided Design*, Vol. 11, No. 1, January 1979, pp. 13–18.

Goodman, T.N.T. "Properties of Beta-splines", *Journal of Approximation Theory*, Vol. 44, No. 2, June 1985, pp. 132–153.

Gordon, William J. "Spline-Blended Surface Interpolation Through Curve Networks", *Journal of Mathematics and Mechanics*, Vol. 18, No. 10, 1969, pp. 931–952. Also Research Publication GMR-921, General Motors Research Laboratories, September 1969.

Gordon, William J. "Blending-Function Methods of Bivariate and Multivariate Interpolation and Approximation", *SIAM Journal on Numerical Analysis*, Vol. 8, 1971, pp. 158–177.

Gordon, William J. "Construction of Curvilinear Co-ordinate Systems and Applications to Mesh Generation", *International Journal for Numerical Methods in Engineering*, Vol. 7, 1973, pp. 461–477.

Gordon, William J. and Riesenfeld, Richard F. "B-spline Curves and Surfaces". In *Computer Aided Geometric Design*, edited by Barnhill, Robert E. and Riesenfeld, Richard F., Academic Press, New York, USA, 1974, pp. 95–126.

Gordon, William J. and Riesenfeld, Richard F. "Bernstein-Bezier Methods for the Computer-Aided Design of Free Form Curves and Surfaces", *Journal of the ACM*, Vol. 21, No. 2, April 1974, pp. 293–310. Also Research Publication GMR-1176, General Motors Research Laboratories, March 1972.

Gould, S.S. "Surface Programs for Numerical Control". In *Proceedings of Curved Surfaces in Engineering*, Churchill College, edited by Brown, I.J., IPC Science and Technology Press, Cambridge, England, 15–17 March 1972.

Gouraud, Henri. *Computer Display of Curved Surfaces*, Ph.D. Thesis, University of Utah, Salt Lake City, Utah, USA, June 1971. Also Tech. Report No. UTEC-CSc-71-113, Department of Computer Science, University of Utah and in *IEEE Transactions on Computers*, Vol. C-20, June 1971, pp. 623–629.

Green, P.J. and Sibson, R. "Computing Dirichlet Tessellations in the Plane", *The Computer Journal*, Vol. 21, No. 2, April 1978, pp. 168–173.

Gregory, John A. *Symmetric Smooth Interpolation on Triangles*, Tech. Report No. T/34, Brunel University, Uxbridge, Middlesex, England, 1973.

Gregory, John A. "Smooth Interpolation Without Twist Constraints". In *Computer Aided Geometric Design*, edited by Barnhill, Robert E. and Riesenfeld, Richard F., Academic Press, New York, USA, 1974, pp. 71–87.

Gregory, John A. *Piecewise Interpolation Theory for Functions of Two Variables*, Ph.D. Thesis, Brunel University, Uxbridge, Middlesex, England, 1975.

Gregory, John A. "A $C^1$ Triangular Interpolation Patch for Computer-Aided Geometric Design", *Computer Graphics and Image Processing*, Vol. 13, No. 1, May 1980, pp. 80–87.

Gross, Jonathan R.; DeRose, Tony D.; and Barsky, Brian A. "Asterisk*: An Extensible testbed for Spline Development", in *Proceedings of Graphics Interface '86*, pp. 241–246, Vancouver, 26–30 May 1986.

Gylys, V.B. "Evaluation of Sculptured Surface Techniques", IIT Research Institute, 1967.

Haber, Robert; Abel, John F.; and Greenberg, Donald P. "A Computer-Aided Design System for Funicular Network Structures". In *Proceedings of the Third International Conference and Exhibition on Computers in Engineering and Building Design*, IPC Science and Technology Press, Guildford, England, 1978, pp. 212–222.

Haber, Robert; Mutryn, Thomas A.; Abel, John F.; and Greenberg, Donald P. "Computer-Aided Design of Framed Dome Structures with Interactive Graphics", *Computer-Aided Design*, Vol. 9, No. 3, July 1977, pp. 157–164.

Haber, Robert; Shepard, Mark; Abel, John F.; Gallagher, Richard H.; and Greenberg, Donald P. "A Generalized Graphic Preprocessor for Two-Dimensional Finite Element Analysis". In *Proceedings of SIGGRAPH '78*, ACM, Atlanta, Georgia, USA, 23–25 August 1978, pp. 323–329.

Hamilton, M.L. and Weiss, A.D. *An Approach to Computer-Aided Preliminary Ship Design*, Tech. Report No. EWSL-TM-228, MIT, Cambridge, Massachusetts, USA, January 1965.

Hanson, Richard J. *Constrained Least Squares Curve Fitting to Discrete Data Using B-splines: A User's Guide*, Numerical Mathematics Division, Sandia Laboratories, Albuquerque, New Mexico, USA, February 1979.

Harris, M. *SPLIFIT: An Interactive Graphics Curve Fitting Computer Program Utilizing Spline Functions*, Report No. PA-TR-4677, Picatinny Arsenal, Dover, New Jersey, USA, January 1974.

Hart, W.B. "Glider Fuselage Design with the Aid of Computer Graphics", *Computer-Aided Design*, Vol. 3, No. 2, 1971, pp. 3–8.

Hart, W.B. "Current and Potential Applications to Industrial Design and Manufacture". In *Proceedings of Curved Surfaces in Engineering*, Churchill College, edited by Brown, I.J., IPC Science and Technology Press, Cambridge, England, 15–17 March 1972.

Hart, W.B. "The Application of Computer-Aided Design Techniques to Glassware and Mould Design", *Computer-Aided Design*, Vol. 4, No. 2, 1972, pp. 57–66.

Hartley, P.J. and Judd, C.J. "Parametrization of Bezier-Type B-spline Curves and Surfaces", *Computer-Aided Design*, Vol. 10, No. 2, March 1978, pp. 130–134.

Hartley, P.J. and Judd, C.J. "Parametrization and Shape of B-spline Curves for CAD", *Computer-Aided Design*, Vol. 12, No. 5, September 1980, pp. 235–238.

Hartley, P.J. and Judd, C.J. "Curve and Surface Representations for Bezier B-spline Systems". In *Proceedings of the CAD80 Conference*, IPC Science and Technology Press, Guildford, England, March 1980, pp. 226–236.

Hatvany, Joseph; Newman, William M.; and Sabin, Malcolm A. "World Survey of Computer-Aided Design", *Computer-Aided Design*, Vol. 9, No. 2, April 1977, pp. 79–98.

Hayes, Jeffrey G. *Available Algorithms for Curve and Surface Fitting*, Report No. NPL-DNACS-39, Division of Numerical Analysis and Computing, National Physical Laboratory, Teddington, Middlesex, England, 1973.

Hayes, Jeffrey G. *New Shapes from Bicubic Splines*, Report No. V. 58, Division of Numerical Analysis and Computing. National Physical Laboratory, Teddington, Middlesex, England, September 1974.

Hayes, Jeffrey G. *Data-Fitting Algorithms Available, in Preparation, and in Prospect for the NAC Library*, Report No. NPL-DNACS-5/78, Division of Numerical Analysis and Computing, National Physical Laboratory, Teddington, Middlesex, England, 1978.

Hayes, J.G. and Halliday, J. "The Least-Squares Fitting of Cubic Spline Surfaces to General Data Sets", *J. Inst. Maths. Applics.*, Vol. 14, 1974, pp. 89–103.

Hazony, Y. "Algorithms for Parallel Processing Curve and Surface Definition with Q-splines", *Computers and Graphics*, Vol. 4, Nos. 3 and 4, 1979, pp. 165–76.

Herron, Gary J. *Triangular and Multisided Patch Schemes*, Ph.D. Thesis, University of Utah, Salt Lake City, Utah, USA, 1979.

Hinds, J.K. and Kuan, L.P. "Surfaces Defined by Curve Transformations". In *Proceedings of the Fifteenth Numerical Control Society Annual Meeting and Technical Conference*, April 1978, pp. 325–340.

Horn, Bertram. "Circle Generators for Display Devices", *Computer Graphics and Image Processing*, Vol. 5, 1976, pp. 280–288.

Hosaka, M. and Kimura, F. "Synthesis Methods of Curves and Surfaces in Interactive CAD". In *Proceedings of the International Conference Interactive Techniques in Computer Aided Design*, IEEE Computer Society, Bologna, Italy, 21–23 September 1978, pp. 151–156.

Hughes, G. and Hankins, H. "Precision Hardware Circular-Arc Generation for Computer Graphic-Display Systems Using Line Segment-Data", *Proceedings of the IEEE*, Vol. 120, No. 2, February 1973, pp. 206–212.

Ichida, K.; Yoshimoto, F.; and Kiyono, T. "Curve Fitting by a One Pass Method with a Piecewise Cubic Polynomial", *ACM Transactions on Mathematical Software*, Vol. 3, 1977, pp. 164–174.

Jiachang, S. "A Class of Matrix Methods for Surface Representation". In *Proceedings of the CAD80 Conference*, IPC Science and Technology Press, Guildford, England, March 1980, pp. 251–254.

Johnson, W.L.; Sanders, J.W.; and South, N.E. "Analytic Surfaces for Computer-Aided Design, SAE Paper 660152". In *Society of Automotive Engineers Congress*, Detroit, Michigan, USA, 1966.

Jordan, B.W.; Lennon, W.J.; and Holm, B.C. "An Improved Algorithm for the Generation of Nonparametric Curves", *IEEE Transactions on Computers*, Vol. C-22, No. 12, December 1973, pp. 1052–1060.

Kay, Douglas S. *Transparency, Refraction, and Ray Tracing for Computer Synthesized Images*, Master's Thesis, Cornell University, Ithaca, New York, USA, January 1979.

Kay, Douglas S. and Greenberg, Donald P. "Transparency for Computer Synthesized Images". In *Proceedings of SIGGRAPH '79*, ACM, Chicago, Illinois, USA, 8–10 August 1979, pp. 158–164.

Klucewicz, I. Michele. "A Piecewise $C^1$ Interpolant to Arbitrarily Spaced Data", *Computer Graphics and Image Processing*, Vol. 8, August 1978, pp. 92–112. Also Master's Thesis, University of Utah, Salt Lake City, Utah, USA, March 1977.

Knapp, Lewis C. *A Design Scheme Using Coons Surfaces with Nonuniform B-spline Curves*, Ph.D. Thesis, Syracuse University, Syracuse, New York, USA, September 1979.

Kojima, T. *Some Geometric Relations Between the Cubic B-spline Curve and Its Associated Characteristic Polygon*, Memo No. CGP78/2, School of Computing Studies and Accountancy, Computational Geometry Project, University of East Anglia, Norwich, England, June 1978.

Kojima, T. "Representation of a Curve Using a Spline Function", *Trans. Soc. Instrum. and Control Eng. (Japan)*, Vol. 14, No. 2, April 1978, pp. 144–148.

Kosugi, M. "Local Curve Fitting Procedures Using Circular Arcs", *Trans. Electron. and Commun. Eng. Jpn. Sect. E (Japan)*, Vol. E60, No. 12, December 1977, pp. 760–761.

Kosugi, M. and Teranishi, T. "Construction of a Curve Segment with Two Circular Arcs", *Trans. Electron. and Commun. Eng. Jpn. Sect. E (Japan)*, Vol. E60, No. 11, November 1977, p. 684.

Kuo, C. *Computer Methods for Ship Surface Design*, Longman, 1971.

Lane, Jeffrey M. *Shape Operators for Computer Aided Geometric Design*, Ph.D. Thesis, University of Utah, Salt Lake City, Utah, USA, June 1977.

Lane, Jeffrey M. and Carpenter, Loren C. "A Generalized Scan Line Algorithm for the Computer Display of Curved Surfaces", *Computer Graphics and Image Processing*, Vol. 11, No. 3, November 1979, pp. 290–297.

Lane, Jeffrey M.; Carpenter, Loren C.; Whitted, J. Turner; and Blinn, James F. "Scan Line Methods for Displaying Parametrically Defined Surfaces", *Communications of the ACM*, Vol. 23, No. 1, January 1980, pp. 23–34.

Lane, Jeffrey M. and Riesenfeld, Richard F. *The Application of Total Positivity to Computer Aided Curve and Surface Design*, Tech. Report No. UUCS-79-115, Department of Computer Science, University of Utah, Salt Lake City, Utah, USA, October 1979.

Lane, Jeffrey M. and Riesenfeld, Richard F. "A Theoretical Development for the Computer Generation of Piecewise Polynomial Surfaces", *IEEE Transactions on Pattern Analysis and Machine Intelligence*, Vol. PAMI-2, No. 1, January 1980, pp. 35–46.

Lane, Jeffrey M. and Riesenfeld, Richard F. "A Geometric Proof of the Variation Diminishing Property of B-spline Approximation", *Journal of Approximation Theory*. Vol. 37, No. 1, January 1983, pp. 1–4.

Lavick, J.J. "Computer-Aided Design at McDonnell Douglas". In *Advanced Computer Graphics–Economics, Techniques and Applications*, edited by Parslow, R.D. and Green, R.E., Plenum Press, London and New York, 1971.

Lawson, C.L. *Generation of a Triangular Grid with Application to Contour Plotting*, Technical Memorandum No. 299, Jet Propulsion Laboratory, Pasadena, California, USA, 1972.

Lawson, C.L. *$C^1$ Compatible Interpolation Over a Triangle*, Computing Memorandum No. 407, Jet Propulsion Laboratory, Pasadena, California, USA, 1976.

Lawson, C.L. "Software for $C^1$ Surface Interpolation". In *Mathematical Software III*, edited by Rice, John R., Academic Press, New York, 1977, pp. 161–194.

Lee, T.M.P. "A Class of Surfaces for Computer Display". In *Proceedings of the Spring Joint Computer Conference*, AFIPS Press, Montvale, New Jersey, USA, 1969, pp. 309–319.

Lee, T.M.P. *Three Dimensional Curves and Surfaces for Rapid Computer Display*, Tech. Report No. ESD-TR-69-189, Dept. Eng. Appl. Phys., Harvard University, Cambridge, Massachusetts, USA, April 1969.

Lee, T.M.P. "Analysis of an Efficient Homogeneous Tensor Representation of Surfaces for Computer Display". In *Advanced Computer Graphics–Economics, Techniques and Applications*, edited by Parslow, R.D. and Green, R.E., Plenum Press, New York, 1971, pp. 195–217.

Levin, Joshua Z. "A Parametric Algorithm for Drawing Pictures of Solid Objects Composed of Quadric Surfaces". *Communications of the ACM*, Vol. 19, No. 10, October 1976, pp. 555–563.

Levin, Joshua Z. "Mathematical Models for Determining the Intersections of Quadric Surfaces", *Computer Graphics and Image Processing*, Vol. 11, September 1979, pp. 73–87.

Levin, Joshua Z. "QUADRIL: A Computer Language for the Description of Quadric-Surface Bodies". In *Proceedings of SIGGRAPH '80*, ACM, Seattle, Washington, USA, 14–18 July 1980, pp. 86–92.

Lewis, B.A. and Robinson, J.S. "Triangulation of Planar Regions with Applications", *The Computer Journal*, Vol. 21, No. 4, 1978, pp. 324–332.

Lillehagen, Frank M.; Riesenfeld, Richard F.; and Frogner, Sverre. "New Dimensions in Man-Machine Communications", *Proceedings of the 3rd International Conference on Computer Applications in the Automation of Shipyard Operation*, Glasgow, Scotland, 1979. Selected for reprint in *Computers in Industry*, Vol. 1, No. 3, 1980, pp. 131–140.

MacCallum, Kenneth J. *The Application of Computer Graphics to the Preliminary Design of Ship Hulls*, Ph.D. Thesis, University of London, England, 1970. Also in Parslow, R.D. and Green, R.E. (editors), *Advanced Computer Graphics–Economics, Techniques and Applications*, Plenum Press, London and New York, 1971, pp. 1203–1216.

MacCallum, Kenneth J. "Surfaces for Interactive Graphical Design", *The Computer Journal*, Vol. 13, No. 4, November 1970, pp. 352–358.

MacCallum, Kenneth J. "Mathematical Design of Hull Surfaces", *The Naval Architect, Journal of the Royal Institute of Naval Architects*, July 1972, pp. 359–373.

Manning, J.R. *Computer-Aided Footwear Design: A Method of Constructing Smooth Curves*, SATRA (Shoe & Allied Trades Research Association) Research Report R.R. No. 251, SATRA House, Rockingham Road, Kettering, Northants., England, December 1972.

Manning, J.R. "Continuity Conditions for Spline Curves", *The Computer Journal*, Vol. 17, No. 2, May 1974, pp. 181–186.

Manning, J.R. "Computerized Pattern Cutting", *Computer-Aided Design*, Vol. 12, No. 1, January 1980, pp. 43–47.

Mansfield, Lois E. "Higher Order Compatible Triangular Finite Elements", *Numerische Mathematik*, Vol. 22, 1974, pp. 89–97.

Mansfield, Lois E. "A Clough-Tocher Type Element Useful for Fourth Order Problems over Nonpolygonal Domains", *Mathematics of Computation*, Vol. 32, 1978, pp. 135–142.

Mansfield, Lois E. "Approximation of the Boundary in the Finite Element Solution of Fourth Order Problems", *SIAM Journal on Numerical Analysis*, Vol. 15, 1978, pp. 568–579.

Marks, R.E. *Surface Fitting for Coons Patches*, CAD Group Document No. 46, Computer Aided Design Group, Cambridge University, Cambridge, England, October 1970.

Marlow, S. and Powell, M.J.D. *A FORTRAN Subroutine for Plotting a Cubic Spline Function*, Report No. AERE-R-7470, UKAEA, Harwell, Berks., England, July 1973.

Marshall, J.A. *Application of Blending Function Methods in the Finite Element Method*, Ph.D. Thesis, University of Dundee, Scotland, 1975.

Marshall, J.A. and Mitchell, A.R. "Blending Interpolants in the Finite Element Method", *International Journal for Numerical Methods in Engineering*, Vol. 12, 1978, pp. 77–83.

Maude, A.D. "Interpolation–Mainly for Graph Plotters", *The Computer Journal*, Vol. 16, 1974, pp. 64–65.

McDermott, Robert J. *Geometric Modelling in Computer Aided Design*, Ph.D. Thesis, University of Utah, Salt Lake City, Utah, USA, March 1980.

McLain, D.H. "Drawing Contours from Arbitrary Data Points", *The Computer Journal*, Vol. 17, 1974, pp. 318–324.

Mehlum, Even. "A Curve-Fitting Method Based on a Variational Criterion, *BIT*, Vol. 4, 1964, pp. 213–223.

Mehlum, Even. "Curve and Surface Fitting Based on a Variational Criterion for Smoothness", Central Institute for Industrial Research (CIIR), Oslo, Norway, 24 December 1969.

Mehlum, Even. "Nonlinear Splines". In *Computer Aided Geometric Design*, edited by Barnhill, Robert E. and Riesenfeld, Richard F., Academic Press, New York, 1974, pp. 173–205.

Mehlum, E. and Sorenson, P.F. "Example of an Existing System in the Shipbuilding Industry: The AUTOKON System", *Proceedings of the Royal Society of London*, Vol. A321, 1971, pp. 219–233.

Moore, C.L. *Method of Fitting a Smooth Surface to a Mesh of Points*, Technical Information Series Report No. 59FPD927, General Electric Company, 1959.

Moore, I.G. "Techniques in Curve Tracing For Computer-Aided Design". In *Computing Australia '74*, Vol. II, Australian Computer Society, Sydney, Australia, 20–24 May 1974, pp. 697–713.

Munchmeyer, Fredrick C. "On the Interactive Design of Smooth Patched Surfaces". In *Proceedings of the International Conference Interactive Techniques in Computer Aided Design*, IEEE Computer Society, Bologna, Italy, 21–23 September 1978, pp. 393–403.

Munchmeyer, Fredrick C.; Schubert, Christian; and Nowacki, Horst. "Interactive Design of Fair Hull Surfaces Using B-splines", *Computers in Industry*, Vol. 1, No. 2, December 1979, pp. 77–86.

Newman, William M. and Sproull, Robert F. *Principles of Interactive Computer Graphics*, McGraw-Hill, 1979. Second edition.

Nicolo, V. *A Preliminary Study for the Choice of the Algorithm to be Used in a 3-D Shape Description Language*, CAD Group Document No. 67, Computer Aided Design Group, Cambridge University, Cambridge, England, August 1972.

Nielson, Gregory M. *Surface Approximation and Data Smoothing Using Generalized Spline Functions*, Ph.D. Thesis, University of Utah, Salt Lake City, Utah, USA, 1970.

Nielson, Gregory M. *Methods for Constructing Combined Smoothing and Interpolating Spline Functions*, Tech. Memo No. 9, Department of Mathematics, Arizona State University, September 1973.

Nielson, Gregory M. "Some Piecewise Polynomial Alternatives to Splines Under Tension". In *Computer Aided Geometric Design*, edited by Barnhill, Robert E. and Riesenfeld, Richard F., Academic Press, New York, 1974, pp. 209–235.

Nielson, Gregory M. *Computation of v-splines*, Tech. Report No. 044-433-11, Department of Mathematics, Arizona State University, June 1974.

Nielson, Gregory M. "Multivariate Smoothing and Interpolating Splines", *SIAM Journal on Numerical Analysis*, Vol. 11, No. 2, April 1974, pp. 435.

Nielson, Gregory M.; Riesenfeld, Richard F.; and Weiss, N.A. "Iterates of Markov Operators", *Journal of Approximation Theory*, Vol. 17, No. 4, August 1976, pp. 321–331.

Notestine, R.E. "Graphics and Computer-Aided Design in Aerospace". In *Proceedings of the National Computer Conference Exposition*, Vol. 42, AFIPS Press, Montvale, New Jersey, USA, 1973, pp. 629–633.

Nowacki, Horst. "Curve and Surface Generation and Fairing". In *Computer Aided Design, Lecture Notes on Computer Science No. 89*, edited by Encarnacao, José L., Springer-Verlag, New York, 1980.

Nutbourne, A.W.; McLellan, P.M. and Kensit, R.M.L. "Curvature Profiles for Plane Curves", *Computer-Aided Design*, Vol. 4, No. 4, 1972, pp. 176–184.

Nydegger, Robert W. *A Data Minimization Algorithm of Analytical Models for Computer Graphics*, Master's Thesis, University of Utah, Salt Lake City, Utah, USA, 1972.

Overhauser, Albert W. *Analytic Definition of Curves and Surfaces by Parabolic Blending*, Tech. Report No. SL68-40, Scientific Research Staff Publications, Ford Motor Company, May 1968.

Pal, T.K. "Intrinsic Spline Curve with Local Control", *Computer-Aided Design*, Vol. 10, No. 1, January 1978, pp. 19–29.

Pal, T.K. "Mean Tangent Rotational Angles and Curvature Integration", *Computer-Aided Design*, Vol. 10, No. 1, January 1978, pp. 30–34.

Pal, T.K. "Hybrid Surface Patch", *Computer-Aided Design*, Vol. 12, No. 6, 1980, pp. 283–287.

Pal, T.K. and Nutbourne, A.W. "Two-Dimensional Curve Synthesis Using Linear Curvature Elements", *Computer-Aided Design*, Vol. 9, No. 2, 1977, pp. 121–134.

Payne, P.J. *A Contouring Program for Joined Surface Patches*, CAD Group Document No. 58, Computer Aided Design Group, Cambridge University, Cambridge, England, June 1971.

Peters, George J. "Interactive Computer Graphics Application of the Parametric Bicubic Surface to Engineering Design Problems". In *Computer Aided Geometric Design*, edited by Barnhill, Robert E. and Riesenfeld, Richard F., Academic Press, New York, 1974. Also in *Proceedings of the National Computer Conference*, AFIPS, 1974, pp. 491–511.

Phong, Bui-Tuong. *Illumination for Computer Generated Images*, Ph.D. Thesis, University of Utah, Salt Lake City, Utah, USA, July 1973. Also Tech. Report No. UTEC-CSc-73-129, Department of Computer Science, University of Utah.

Phong, Bui-Tuong, "Illumination for Computer Generated Pictures", *Communications of the ACM*, Vol. 18, No. 6, June 1975, pp. 311–317.

Pilcher, David T. *Smooth Approximation of Parametric Curves and Surfaces*, Ph.D. Thesis, University of Utah, Salt Lake City, Utah, USA, August 1973.

Pitteway, M.L.V. "Algorithm for Drawing Ellipses or Hyperbolae with a Digital Plotter", *The Computer Journal*, Vol. 10, No. 3, November 1967, pp. 282–289.

Pitteway, M.L.V. "Integer Circles, etc.–Some Further Thoughts", *Computer Graphics and Image Processing*, Vol. 3, 1974, pp. 262–265.

Pitteway, M.L.V. and Botting, R. "Integer Circles, etc.–Three More Extensions of Bresenham's Algorithm", *Computer Graphics and Image Processing*, Vol. 3, 1974, pp. 260–261.

Pitteway, M.L.V. and Waltkinson, D.J. "Bresenham's Algorithm with Grey Scale", *Communications of the ACM*, Vol. 23, No. 11, November 1980, pp. 625–626. Corrigendum in *Communications of the ACM*, Vol. 24, No. 2, February 1981, p. 88.

Poeppelmeier, Charles C. *A Boolean Sum Interpolation Scheme to Random Data for Computer Aided Geometric Design*, Master's Thesis, University of Utah, Salt Lake City, Utah, USA, 1975.

Porter, Thomas K. "Spherical Shading". In *Proceedings of SIGGRAPH '78*, ACM, Atlanta, Georgia, USA, 23–25 August 1978, pp. 282–285.

PorterGoff, R.F.D. *The Representation of Surfaces by Coons' Patch Technique*, Tech. Report No. 69-3, University of Leicester, England, February 1969.

Ramer, U. "An Iterative Procedure for the Polygonal Approximation of Plane Curves", *Computer Graphics and Image Processing*, Vol. 1, No. 3, November 1972, pp. 244–256.

Ramot, J. "Nonparametric Curves", *IEEE Transactions on Computers*, Vol. C-25, No. 1, January 1976, p. 103.

Rice, John R. "Algorithm 525: ADAPT, Adaptive Smooth Curve Fitting (E2)", *ACM Transactions on Mathematical Software*, Vol. 4, No. 1, March 1978, pp. 82–94.

Riesenfeld, Richard F. *Applications of B-spline Approximation to Geometric Problems of Computer-Aided Design*, Ph.D. Thesis, Syracuse University, Syracuse, New York, USA, May

1973. Available as Tech. Report No. UTEC-CSc-73-126, Department of Computer Science, University of Utah, Salt Lake City, Utah, USA.

Riesenfeld, Richard F. "Nonuniform B-spline Curves". In *Proceedings of the 2nd USA-Japan Computer Conference*, AFIPS, Montvale, New Jersey, USA, 26–28 August 1975, pp. 551–555.

Riesenfeld, Richard F. "Aspects of Modelling in Computer Aided Geometric Design". In *Proceedings of the National Computer Conference*, AFIPS, May 1975, pp. 597–602.

Riesenfeld, Richard F. "On Chaikin's Algorithm", *Computer Graphics and Image Processing*, Vol. 4, No. 3, September 1975, pp. 304–310.

Riesenfeld, Richard F. "Realtime Computer Graphics in Interactive Curve and Surface Design". In *Proceedings of the Fall Company Conference*, IEEE Computer Society, 1976, pp. 122–124.

Riesenfeld, Richard F. "Computer-Aided Design in Japan's Shipbuilding Industry", *Office of Naval Research Scientific Bulletin*, Vol. 3, No. 3, 1978, pp. 43–48.

Riesenfeld, Richard F.; Cohen, Elaine; Fish, Russell D.; Thomas, Spencer W.; Cobb, Elizabeth S.; Barsky, Brian A.; Schweitzer, Dino L.; and Lane, Jeffrey M. "Using the Oslo Algorithm as a Basis for CAD/CAM Geometric Modelling". In *Proceedings of the Second Annual NCGA National Conference*, National Computer Graphics Association, Inc., Baltimore, Maryland, USA, 14–18 June 1981, pp. 345–356.

Roberts, Larry G. "Conic Display Generator Using Multiplying Digital-Analog Converters", *IEEE Transactions on Electronics and Computers*, Vol. EC-16, No. 3, June 1967, p. 369.

Rogers, David F. and Adams, J. Alan. *Mathematical Elements for Computer Graphics*, McGraw-Hill, New York, USA, 1976.

Rogers, David F. and Satterfield, Steven G. "B-spline Surfaces for Ship Hull Design". In *Proceedings of SIGGRAPH '80*, ACM, Seattle, Washington, USA, 14–18 July 1980, pp. 211–217.

Rubin, F. "Generation of Nonparametric Curves", *IEEE Transactions on Computers*, Vol. C-25, No. 1, January 1976, p. 103.

Rubin, Steven M. and Whitted, J. Turner. "A 3-Dimensional Representation for Fast Rendering of Complex Scenes". In *Proceedings of SIGGRAPH '80*, ACM, Seattle, Washington, USA, 14–18 July 1980, pp. 110–116.

Sabin, Malcolm A. *Spline Surfaces*, Report No. VTO/MS/156, Dynamics and Mathematical Services Dept., British Aircraft Corporation, Weybridge, Surrey, England, June 1968.

Sabin, Malcolm A. *Parametric Surface Equations for Non-rectangular Regions*, Report No. VTO/MS/147, Dynamics and Mathematical Services Dept., British Aircraft Corporation, Weybridge, Surrey, England, July 1968.

Sabin, Malcolm A. *Numerical Master Geometry*, Report No. VTO/MS/146, Dynamics and Mathematical Services Dept., British Aircraft Corporation, Weybridge, Surrey, England August 1968.

Sabin, Malcolm A. *Offset Parametric Surfaces*, Report No. VTO/MS/149, Dynamics and Mathematical Services Dept., British Aircraft Corporation, Weybridge, Surrey, England, September 1968.

Sabin, Malcolm A. *Two Basic Interrogations of Parametric Surfaces*, Report No. VTO/MS/148, Dynamics and Mathematical Services Dept., British Aircraft Corporation, Weybridge, Surrey, England, October 1968.

Sabin, Malcolm A. *General Interrogations of Parametric Surfaces*, Report No. VTO/MS/150, Dynamics and Mathematical Services Dept., British Aircraft Corporation, Weybridge, Surrey, England, October 1968.

Sabin, Malcolm A. *Conditions for Continuity of Surface Normal between Adjacent Parametric Surfaces*, Report No. VTO/MS/151, Dynamics and Mathematical Services Dept., British Aircraft Corporation, Weybridge, Surrey, England, October 1968.

Sabin, Malcolm A. *The Use of Vectors and Parameters to Describe Geometrical Concepts*, Report No. VTO/MS/152, Dynamics and Mathematical Services Dept., British Aircraft Corporation, Weybridge, Surrey, England, December 1968.

Sabin, Malcolm A. *A 16-Point Bicubic Formulation Suitable for Multipatch Surfaces*, Report No. VTO/MS/155, Dynamics and Mathematical Services Dept., British Aircraft Corporation, Weybridge, Surrey, England, March 1969.

Sabin, Malcolm A. *Parametric Splines in Tension*, Report No. VTO/MS/160, British Aircraft Corporation, Weybridge, Surrey, England, 23 July 1970.

Sabin, Malcolm A. "An Existing System in the Aircraft Industry. The British Aircraft Corporation Numerical Master Geometry System". *Proceedings of the Royal Society of London*, Vol. A321, 1971, pp. 197–205.

Sabin, Malcolm A. "Interrogation Techniques for Parametric Surfaces". In *Advanced Computer Graphics–Economics, Techniques and Applications*, edited by Parslow, R.D. and Green, R.E., Plenum Press, London and New York, 1971, pp. 1095–1118.

Sabin, Malcolm A. "Comments on Some Algorithms for the Representation of Curves by Straight Line Segments", Letter in *The Computer Journal*, Vol. 15, No. 2, 1972, p. 104.

Sabin, Malcolm A. "A Method for Displaying the Intersection Curve of Two Quadric Surfaces", *The Computer Journal*, Vol. 19, November 1976, pp. 336–338.

Sabin, Malcolm A. "Cursive Script Output", *Software–Practice and Experience*, Vol. 6, October/December 1976, p. 581.

Sabin, Malcolm A. *The Use of Piecewise Forms for the Numerical Representation of Shape*, Ph.D. Thesis, Budapest, Hungary, 1976.

Sablonniere, P. "Spline and Bézier Polygons Associated with a Polynomial Spline Curve", *Computer-Aided Design*, Vol. 10, No. 4, July 1978, pp. 257–261.

Satterfield, Steven G.; Rodriguez, Francisco; and Rogers, David F. "A Simple Approach to Computer Aided Milling with Interactive Graphics". In *Proceedings of SIGGRAPH '77*, ACM, San José, California, USA, 20–22 July 1977, pp. 107–111.

Schechter, A. "Synthesis of 2D Curves by Blending Piecewise Linear Curvature Profiles", *Computer-Aided Design*, Vol. 10, No. 1, January 1978, pp. 8–18.

Schechter, A. "Linear Blending of Curvature Profiles", *Computer-Aided Design*, Vol. 10, No. 2, March 1978, pp. 101–109.

Schmitt, Francis J.M.; Barsky, Brian A.; and Du, Wen-Hui. "An Adaptive Subdivision Method for Surface Fitting from Sampled Data", in *SIGGRAPH '86 Conference Proceedings*, Vol. 20, pp. 179–188, ACM, Dallas, August 18–22, 1986.

Schoenberg, I.J. and Whitney, A. "On Polya Frequency Functions III: The Positivity of Translation Determinants with an Application to the Interpolation Problem by Spline Curves", *Transactions of the American Mathematical Society*, Vol. 74, 1953, pp. 246–259.

Schumaker, Larry L. "Fitting Surfaces to Scattered Data". In *Approximation Theory II*, edited by Lorentz, G.G.; Chui, C.K.; and Schumaker, L.L., Academic Press, New York, USA, 1976, pp. 203–268.

Schweikert, Daniel G. "An Interpolation Curve Using a Spline in Tension", *Journal of Mathematics and Physics*, Vol. 45, 1966, pp. 312–317.

Shapira, R. and Freeman, H. "Computer Description of Bodies Bounded by Quadric Surfaces from a Set of Imperfect Bodies", *IEEE Transactions on Computers*, Vol. C-27, September 1978, pp. 841–854.

Shephard, G.B. *Analytic Approximations to Smooth Contour Lines of Vehicle Body Panels*, CAD Group Document No. 55, Computer Aided Design Group, Cambridge University, Cambridge, England, December 1970.

Shu, H.; Hori, S.; Mann, W.R.; and Little, R.N. "The Synthesis of Sculptured Surfaces". In *Numerical Control Programming Languages*, edited by Leslie, W.H.P., North-Holland, Amsterdam, Holland, 1970, pp. 358–375.

Sibson, R. "Locally Equiangular Triangulations", *The Computer Journal*, Vol. 21, No. 3, 1978, pp. 243–245.

Smith, D.J.L. and Merryweather, H. "The Use of Analytic Surfaces for the Design of Centrifugal Impellers by Computer Graphics", *International Journal for Numerical Methods in Engineering*, Vol. 17, 1973, pp. 137–154.

Soanes, R.W. Jr. "VP-splines, an Extension of Twice Differentiable Interpolation". In *Proceedings of the 1976 Army Numerical Analysis and Computer Conference (ARO Report 76-3)*, U.S. Army Research Office, 1976, pp. 141–152.

South, N.E. and Kelly, J.P. *Analytic Surface Methods*, Ford Motor Company, N/C Development Unit, Production Engineering Office, December 1965.

Spaeth, H. "Exponential Spline Interpolation", *Computing*, Vol. 4, 1969, pp. 225–233.

Spaeth, H. *Spline-Algorithmen zur Konstruktion glatter Kurven und Flächen*, R. Oldenbourg Verlag, Munich, Germany, 1973. English translation by Hoskins, W.D. and Sager, H.W. as

*Spline Algorithms for Curves and Surfaces*, Utilitas Mathematica Publ., Winnipeg, Manitoba, Canada, 1974.

Stotz, Robert. *Specialized Computer Equipment for Generation of Three-Dimensional Curvilinear Figures*, Tech. Report No. ESL-TM-167, Electron. Systems Lab., MIT, Cambridge, Massachusetts, USA, January 1963.

Strasser, Wolfgang. *Schnelle Kurven- und Flächendarstellung auf grafischen Sichtgeräten*, Ph.D. Thesis, Technische Universität Berlin, Berlin, Germany, September 1974.

Sun, Jia-Chang. "The Spline Interpolation for Space Curves in Local Coordinates", *Acta Math. App. Sin. (China)*, Vol. 2, No. 4, November 1979, pp. 340–343.

Sutcliffe, D. "An Algorithm for Drawing the Curve f(x, y) = 0", *The Computer Journal*, Vol. 19, No. 3, August 1976, pp. 246–249.

Tallbot, J. *Experiments Towards Interactive Graphical Design of Motor Bodies*, CAD Group Document No. 56, Computer Aided Design Group, Cambridge University, Cambridge, England, December 1979.

Thalmann, Daniel and Magnenat-Thalmann, Nadia. *Computer Animation: Theory and Practice*, Springer-Verlag, Berlin, Germany, 1985.

Thomas, D.H. "Pseudospline Interpolation for Space Curves", *Mathematics of Computation*, Vol. 30, No. 133, 1976, pp. 58–67.

Tilove, Robert B. *B-splines for the Uninitiated*, Memo No. CGP78/9, School of Computing Studies and Accountancy, Computational Geometry Project, University of East Anglia, Norwich, England, August 1978.

Timmer, H.G. "Alternative Representation for Parametric Cubic Curves and Surfaces", *Computer-Aided Design*, Vol. 12, No. 1, January 1980, pp. 25–28.

Ts'o, Pauline and Barsky, Brian A. "Modelling and Rendering Waves: Wave-tracing Using Beta-splines and Reflective and Refractive Texture Mapping", *ACM Transactions on Graphics*, Vol. 6, No. 2, April 1987, to appear.

Veenman, P. and den Hartog, C.G. "DIECAST-Display Interaction Enhancing Computer Aided Shape Technique". In *Proceedings of ONLINE 72*, Brunel University, Uxbridge, England, 1972.

Veron, M.; Ris, G.; and Musse, J.-P. "Continuity of Biparametric Surface Patches", *Computer-Aided Design*, Vol. 8, No. 4, 1976, pp. 267–273.

Versprille, Kenneth J. *Computer-Aided Design Applications of the Rational B-spline Approximation Form*, Ph.D. Thesis, Syracuse University, Syracuse, New York, USA, February 1975.

Vittitow, W.L. *Interpolation to Arbitrarily Spaced Data*, Ph.D. Thesis, University of Utah, Salt Lake City, Utah, USA, 1978.

Walker, L.F. "Curved Surfaces in Shipbuilding Design and Production". In *Proceedings of Curved Surfaces in Engineering*, Churchill College, edited by Brown, I.J., IPC Science and Technology Press, Cambridge, England, 15–17 March 1972.

Walter, H, "Computer-Aided Design in the Aircraft Industry". In *Computer-Aided Design*, edited by Vlietstra, J. and Wielinga, R.F., North-Holland, Amsterdam, Holland, 1973.

Weichbrodt, Andreas. *SHIPDS–SHIPLO: A Two Phase Programming System for Surface Representation in Shipbuilding and Engineering*, Master's Thesis, University of Utah, Salt Lake City, Utah, USA, March 1980.

Weiss, Ruth A. "BE VISION, A Package of IBM 7090 FORTRAN Programs to Draw Orthographic Views of Plane and Quadric Surfaces", *Journal of the ACM*, Vol. 13, No. 2, April 1966, pp. 194–204.

Whitted, J. Turner, "An Improved Illumination Model for Shaded Display", *Communications of the ACM*, Vol. 23, No. 6, June 1980, pp. 343–349.

Wielinga, R.F. "Constrained Interpolation Using Bezier Curves as a New Tool in ComputerAided Geometric Design". In *Computer Aided Geometric Design*, edited by Barnhill, Robert E. and Riesenfeld, Richard F., Academic Press, New York, USA, 1974.

Williams, Lance. "Casting Curved Shadows on Curved Surfaces". In *Proceedings of SIGGRAPH '78*, ACM, Atlanta, Georgia, USA, 23–25 August 1978, pp. 270–274.

Woo, T. "Progress in Shape Modelling", *IEEE Computer*, Vol. 10, December 1977, pp. 40–46.

Woon, P.Y. *On the Computer Drawing of Solid Objects Bounded by Quadric Surfaces*, Tech. Report No. TR-403-3, Department of Computer Science, New York University, New York, USA, June 1969.

Woon, P.Y. *A Computer Procedure for Generating Visible Line Drawings for Solids Bounded by Quadric Surfaces*, Tech. Report No. TR-403-15, New York University, New York, USA, December 1970.

Woon, P.Y. and Freeman, H. "A Computer Procedure for Generating Visible Line Drawings for Solids Bounded by Quadric Surfaces". In *Proceedings of the IFIP Congress, Information Processing '71*, North-Holland, Amsterdam, Holland, 1971, pp. 1120–1125.

Wu, Sheng-Chuan and Abel, John F. "Representation and Discretization of Arbitrary Surfaces for Finite Element Shell Analysis", *International Journal for Numerical Methods in Engineering*, Vol. 14, No. 6, 1979, pp. 813–836.

Wu, Sheng-Chuan; Abel, John F.; and Greenberg, Donald P. "An Interactive Computer Graphics Approach to Surface Representation", *Communications of the ACM*, Vol. 20, No. 10, October 1977, pp. 703–712.

Yamaguchi, Fujio. "A Design Method of Free Form Surfaces by a Computer Display–Curve Design and Initial Surface Generation Methods", *J. Jpn. Soc. Precis. Eng. (Japan)*, Vol. 43, No. 9, September 1977, pp. 1105–1111.

Yamaguchi, Fujio. "A New Curve Fitting Method Using a CRT Computer Display", *Computer Graphics and Image Processing*, Vol. 7, No. 3, June 1978, pp. 425–437.

Yamaguchi, Fujio. *A Design System for Free Form Objects* (FREEDOM), Technical Research Institute, Japan Society for the Promotion of Machine Industry, 1976.

Yuille, I.M. "Ship Design". In *Proceedings of Curved Surfaces in Engineering*, Churchill College, edited by Brown, I.J., IPC Science and Technology Press, Cambridge, England, 15–17 March 1972.

# Subject Index

# Computer Science Workbench

Editor: T. L. Kunii

N. Magnenat-Thalmann, D. Thalmann

# Computer Animation

**Theory and Practice**

1985. 156 figures, 54 of them in color. XIII, 240 pages.
ISBN 3-540-70005-6

**Contents:** Introduction. – Conventional Animation. – Computer Animation. – The Development of Computer Animation in Various Organizations. – Key Frame and Painting Systems. – Modeled Animation. – Hidden Surfaces, Reflectance and Shading. – Transparency, Texture, Shadows and Anti-aliasing. – Human Modeling and Animation. – Object-oriented and Actor Languages and Systems. – Case Studies. – A Case Study: *Dream Flight*. – References. – Appendix A–C. – Subject Index.

N. Magnenat-Thalmann, D. Thalmann

# Image Synthesis

**Theory and Practice**

1987. 223 figures, 80 of them in color. XV, 400 pages.
Hard cover DM 148,–. ISBN 3-540-70023-4

The book contains a detailed description of the most fundamental algorithms; other less important algorithms are summarized or simply listed. This volume is also a unique handbook of mathematical formulae for image synthesis. The four first chapters of the book survey the basic techniques of computer graphics which play an important role in the design of an image: geometric models, image and viewing transformations, curves and surfaces and solid modeling techniques. In the next chapters, each major topic in image synthesis is presented.

Springer-Verlag
Berlin Heidelberg New York
London Paris Tokyo